Teaching Physics

with the Physics Suite

EDWARD F. REDISH
University of Maryland

JOHN WILEY & SONS, INC.

ACQUISITIONS EDITOR	Stuart Johnson
MARKETING MANAGER	Robert Smith
ASSOCIATE PRODUCTION MANAGER	Kelly Tavares
DESIGN DIRECTOR	Maddy Lesure

This book was set in 10/12 Adobe Garamond by Matrix Publishing and printed and bound by Malloy Lithographing. The cover was printed by Phoenix Color.

ISBN 10: 0-471-39378-9
ISBN 13: 978-0-471-39378-8

Printed in the United States of America

10 9 8 7 6 5 4 3

PREFACE

A preface is the place where you expect to have some questions answered: What is this book about? Who is the intended audience? What value might I expect to get out of it?

Although this book is associated with the Physics Suite and is distributed as an instructor's guide to using the Physics Suite, it really tries to be more than just a how-to book for a particular set of materials. This book is about learning to be a more effective teacher.

The intended audience is any physics teacher who is interested in learning about recent developments in physics education. In contrast to some other teachers' guides, it is not a review of specific topics in physics with hints on how to teach them and lists of what common student difficulties are. Rather, it is a handbook with a variety of tools for improving both the teaching and learning of physics—from new kinds of homework and exam problems to surveys for figuring out what has happened in your class to tools for taking and analyzing data using computers and video.

Over the past two decades, a community has grown up in the scholarship of teaching and learning, bridging physics and education. The people in this community, which I refer to as *Physics Education Research* (PER), have tried to understand why so many students have so much difficulty understanding physics, and they have tried to develop learning environments to help those students. The result has been a large body of knowledge and a growing repertoire of curricula that are demonstrably more effective than our traditional approaches.

The Physics Suite integrates materials from an active group of PER developers. All the parts of the Physics Suite are based on education research and share a specific underlying philosophy. This book, while providing an introduction to the materials of the Physics Suite, more importantly, provides an introduction to the educational philosophy and knowledge base that form the foundation underlying the Suite. Because this educational philosophy and knowledge base rest on well-documented scholarship, the foundation is broadly applicable. This foundation and this book can help you to teach better even if you do not adopt a single item from the Physics Suite.

Because the character of the book arises in a very personal way from my own experiences, let me introduce myself briefly. I was trained as a theoretical nuclear physicist and began my teaching career at the University of Maryland in 1970. I have been on the Maryland faculty ever since, teaching and doing research. From the first, I had a strong interest in teaching. From the first, I cheerfully ignored my colleagues' advice to put a minimal effort into my teaching duties, since they would play little role in helping me get tenure.

In the 1980s, I worked on trying to get the newly invented personal computer into my classes. But as the decade went on, I became increasingly aware of two important facts: First, that my students were having trouble learning physics—both with and without the computer—and that their problems were more difficult to resolve than I had expected; and second, that there was a community that knew this and was studying it as a research effort. In 1991, I stopped doing nuclear physics and switched my research activity to PER.

This history determines the structure of the book. Some of what I have learned in 30 years of teaching has been from the research literature in PER and from my own work as an active physics education researcher. But a lot has been from listening to and working with the students in my physics classes over all those years. As a result, some of what helps my teaching is well documented through published research, but some is not.

Therefore, I have chosen to present this book neither as a research monograph nor as a standard didactic "how-to" teaching guide. Rather, I have decided to make it a "teacher-to-teacher" discussion in which I present what I have learned in three ways: as research results with data and citations where they are available, as illustrative anecdotal examples from my own experience, and as general principles, guidelines, and heuristics that I have found helpful.

The anecdotal examples do not always follow the events exactly. Sometimes multiple stories have been combined and details omitted for clarity. Although these stories are based on real events, they should be interpreted as fables with morals rather than as records of real events.

I have tried to limit my general principles to those that can be supported in three ways: by observations of real student behavior in real classrooms (usually by educational researchers), by controlled experimental studies on how people think (usually by cognitive scientists), and by physiological plausibility (consistency with what is known in neuroscience). Heuristics (such as "Redish's Teaching Commandments") are less well-documented and are based on my own experience and on what I have learned from other physics teachers.

Throughout my career as a research physicist, my work had a strongly theoretical bias. I have always been interested in trying to understand how to think about and organize our knowledge of the real world. In trying to understand the system of students trying to learn physics, the appropriate theory to help us parse what we see into something sensible is cognitive science. As a result, this book has a strong cognitive flavor. Though my intent is not to write a textbook in cognitive science, I have tried to extract and make plausible for physicists what is relevant and known in this area. For those who want more documentation or want to better understand the strengths and limitations of what is known, I refer you to the references cited in the text.

This book has four parts:

- An introduction discussing the structure of the Physics Suite and the motivation for educational reform in introductory physics. (chapter 1)
- A discussion of what is known about how people think that is relevant for physics teaching and learning. (chapters 2 and 3)
- Two chapters about assessing individual students' learning and evaluating the success of instruction for a class. (chapters 4 and 5)
- A survey of various methods for creating learning environments that can help to improve student learning, including both tips from my own classroom experience and descriptions of the PER-based curricular materials and methods belonging to the Physics Suite and some other methods that work well with it. (chapters 6–10)

Finally, the book comes with a *Resource CD.* This contains

- Our Action Research Kit—a collection of concept and attitude surveys
- Resources for exploring computer-assisted data acquisition and analysis and video data handling
- Resources for getting information about PER

In the Appendix to this volume, I list the material available on the disk. The disk is attached inside the back cover. If it is missing in your copy, contact John Wiley & Sons to get one.

ACKNOWLEDGMENTS

Throughout my studies of PER, a number of individuals have been tremendously helpful, both through their published work and through personal conversations. First and foremost is Lillian C. McDermott, not only through her large and informative body of research, but through taking me in as a sabbatical visitor to her well-established research group in PER at the University of Washington in 1992–1993. This gave me an excellent start in learning how to do PER and a view of what a PER group inside a physics department looked like.

Others whose work had a primary influence on my thinking include the late Arnold Arons, John Clement, Fred Goldberg, David Hammer, Pat Heller, David Hestenes, Jose Mestre, and Fred Reif. I also want to thank the students, postdocs, and visitors I have worked with in PER at Maryland. I have learned much from them, and discussions with them have helped me clarify and refine my thinking on many occasions: in alphabetical order, they are Jonte Bernhard, John Christopher, Andy Elby, Paul Gresser, Apriel Hodari, Beth Hufnagel, Pratibha Jolly, Bao Lei, Rebecca Lippmann, Laura Lising, Tim McCaskey, Seth Rosenberg, Mel Sabella, Al Sapirstein, Jeff Saul, Rachel Scherr, Richard Steinberg, Jonathan Tuminaro, Zuyuan Wang, and Michael Wittmann. Through the past decade, my collaborators in the Activity-Based Physics Group have been invaluable in both helping me develop my views on education and in the creation of this book: Pat Cooney, Karen Cummings, Priscilla Laws, David Sokoloff, and Ron Thornton.

I would like to thank those people who commented on various versions of the text, especially those who helped clarify my descriptions of their work: Bob Beichner, Mary Fehrs, Gary Gladding, Ken and Pat Heller, Paula Heron, Priscilla Laws, Eric Mazur, Lillian McDermott, Evelyn Patterson, David Sokoloff, Ron Thornton, and Maxine Willis. Priscilla Laws and Tim McCaskey did careful readings of my draft and made many valuable suggestions.

I want to acknowledge a grant from the University of Maryland Graduate Research Board that played a major role in allowing me to take a sabbatical to write this book. I would also like to thank the Graduate School of Education at UC Berkeley for hosting that sabbatical, with particular thanks to the following for valuable discussions: Michael Ranney, Andy diSessa, Alan Schoenfeld, and Barbara White. Much of my research that is cited here has been supported by the U.S. National Science Foundation and the Fund for the Improvement of Post-Secondary Education of the U.S. Department of Education.

Finally, special thanks are due to my wife, Janice (Ginny) Redish, not only for her support and encouragement throughout, but for her outstanding skill and expertise in editing and technical communication. She was immensely helpful in making the book more readable. She has also been an invaluable resource in helping me both find and understand what is known and relevant in cognitive science and the study of human behavior.

Edward F. (Joe) Redish
University of Maryland
College Park

Contents

CHAPTER 1

Introduction and Motivation

The Master doesn't talk, he acts.
When his work is done
[his students] say, "Amazing:
we did it, all by ourselves."
Lao Tse, Tao Te Ching [Mitchell 1988]

INTRODUCTION

Teaching physics can be both inspirational and frustrating. Those of us who enjoy learning physics get to rethink and pull together what we know in new and coherent ways. We enjoy the opportunity to create new demonstrations, invent new derivations, and solve interesting problems. For those of us who love doing physics, teaching can be a delightful learning experience. Occasionally we find a student who has the interest and ability to understand what we are trying to do and who is inspired and transformed by our teaching. That makes all the frustrations worthwhile.

On the other hand, there are frustrations. We may have students who seem unable to make sense of what we do—sometimes a lot of them. They are confused and even hostile. We may make intense efforts to reach these students, either by making our classes more entertaining or by simplifying what we ask them to do. While these efforts may lead to better student evaluations, they rarely result in our students understanding more physics. They can lead to a "dumbing down" of the physics that we find frustrating and disappointing.

Can we reduce this frustration and find ways to reach those students who don't seem to "get it"? In the past two decades, there has been a growing understanding of why so many students respond badly to traditional physics instruction and of how to modify our instructional methods to help them learn more. A number of researchers and curriculum developers have begun to weave the results of education research and new technological tools into more effective learning environments.

One result of this interweaving of research and technology is the *Physics Suite*. In the Physics Suite, the Activity-Based Physics (ABP) Group[1] is creating a new kind of educational environment. Since there is a growing diversity of environments for introductory physics, the ABP Group has opted for a modular structure, one that can be implemented a step at a time or adopted in a total makeover, affecting all parts of the course. This book is about how our teaching of physics can change as a result of these new environments. It discusses the elements of this modular structure, how to use them, and the educational philosophy, cognitive theory, pedagogical research, and modern technology on which the Physics Suite is based.

TYPICAL MATERIALS FOR A PHYSICS CLASS

Typically, all the materials offered by a publisher for a physics class derive from the text (see Figure 1.1). Affiliated materials are available associated with the text, usually including everything from a "quick summary" for students to colored transparencies for instructors. There may be a CD with (often uninteresting) simulations that offer little or no guidance for either students or instructors in how to use it to make it pedagogically effective. Adopting institutions may add a laboratory with lessons developed on site. But the text is primary, and its selection usually depends critically on content—what is covered and whether it is treated correctly. Those are certainly important criteria.

But extensive research has shown that effective student learning seldom comes from the text. Students frequently have difficulty making sense of a physics text, and, as a result, only a small minority actually read the text in the careful and thoughtful way we expect. Effective student learning comes from "brains-on" activities—those times when they are thinking hard and struggling to make sense of what they are learning. Effective instruction happens when we create environments in which students are encouraged and helped to engage in those kind of activities. Well-tested innovations that focus on building reasoning through carefully planned and structured *activities* in lectures, recitations, laboratories, or workshops are more

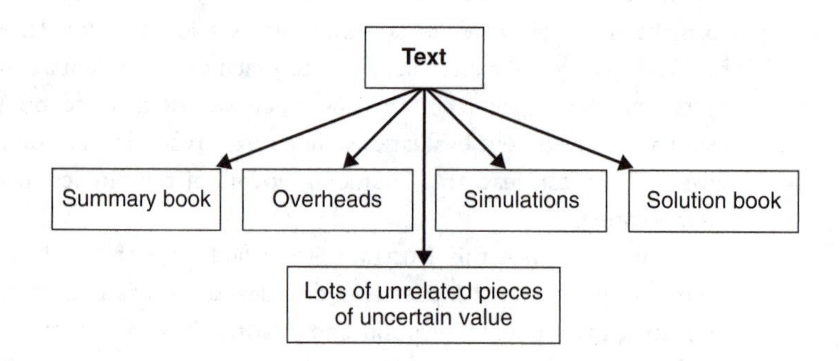

Figure 1.1 Typical layout of materials associated with a physics course.

[1] Pat Cooney, Karen Cummings, Priscilla Laws, David Sokoloff, Ron Thornton, and myself.

likely to produce strong student learning. For most students, these activities play as important a role as does reading the text.

The Physics Suite is much more than a text with a collection of ancillaries developed after the fact. The Physics Suite builds on integrating a series of strong activity-based elements with the text. The Physics Suite focuses on getting students to learn to *do* what they need to do to learn physics.

A NEW ALTERNATIVE: THE PHYSICS SUITE

The ABP Group has created a new structure consisting of a broad array of integrated educational materials: *The Physics Suite.* These materials are shown schematically in Figure 1.2. Two

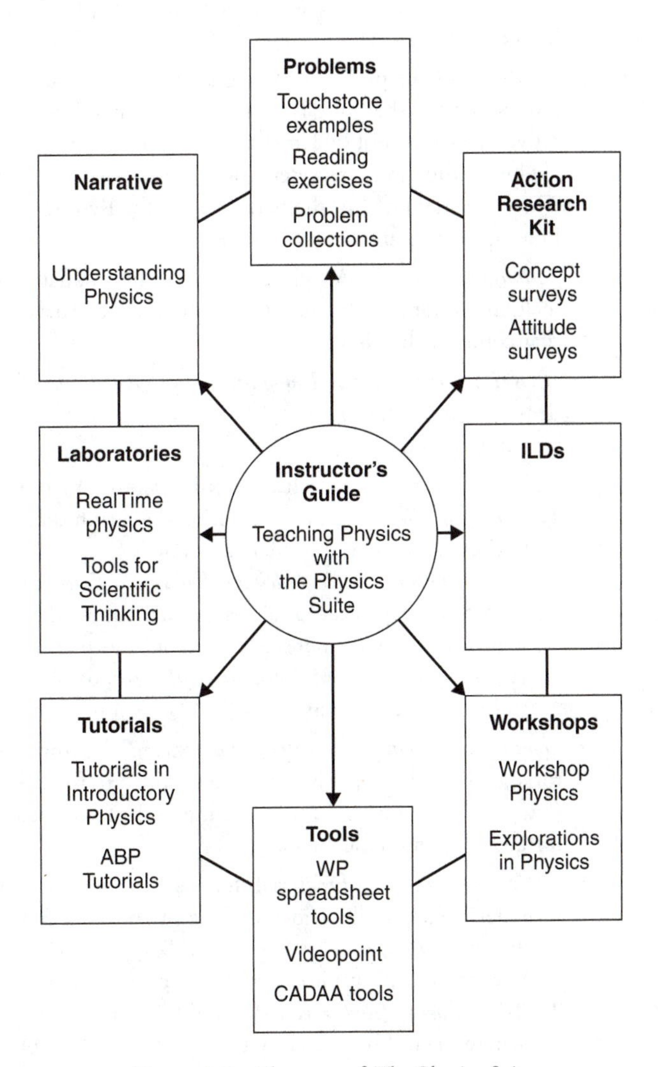

Figure 1.2 Elements of The Physics Suite

particular themes of many Suite elements are: (1) the use of guided activities to help students construct their learning, and (2) the use of modern technology, particularly *computer-assisted data acquisition and analysis* (CADAA). The Physics Suite consists of the following materials:

- *The Instructor's Guide: Teaching Physics with the Physics Suite (Redish)*—This book: a guide not only to the materials of the Suite (and other curricular materials that fit well with the Suite), but a discussion of the motivation, theoretical frame, and data describing its effectiveness.
- *Narrative: Understanding Physics (Cummings, Laws, Redish, and Cooney)*—A revised version of the classic Halliday, Resnik, and Walker text, modified to focus more strongly on issues where students are known to have difficulties from educational research. The new narrative also stresses the empirical base for physics knowledge, explaining not only what we know but how we know it. (See chapter 10 for more details on how the text has been changed.)
- *Problems*—Since problem solving is one of the places where students learn the most in a physics class, the Suite is enriched by a careful choice of sample problems in the narrative (Touchstone Problems) and by supplementary problems of a variety of types including estimations, representation translations, and context-rich real-world problems. These are contained in the narrative, on the Resource CD in the back of this book, and in a supplementary book of problems.
- *Action Research Kit*—A collection of tools for evaluating the progress of one's instruction, including a variety of concept tests and attitude surveys. These are on the Resource CD that comes with this volume.
- *ILDs: Interactive Lecture Demonstrations (Sokoloff and Thornton)*: Worksheet-based guided demonstrations that use CADAA to help students build concepts and learn representation translation.
- *Workshops*—Three sets of full workshop/studio materials are associated with the Suite.
  1. *Workshop Physics (Laws)*: A full lab-based physics course at the calculus level using CADAA, video, and numerical modeling.
  2. *Explorations in Physics (Jackson, Laws, and Franklin)*: Developed as part of the Workshop Science project, a lab-based curriculum that uses less mathematics and is designed for use with nonscience majors and preservice teachers.
  3. *Physics by Inquiry (McDermott et al.)*: A workshop-style course appropriate for pre- and in-service teachers.
- *Tools (Laws, Cooney, Thornton, and Sokoloff)*: Computer tools for use in laboratory, tutorial, and Workshop Physics include software for collecting, displaying, and analyzing CADAA data, software for extracting and plotting data from videos, and spreadsheets for analyzing numerical data.
- *Tutorials*—Two sets of materials for use in recitation sections that use guided small-group activities to help build understanding of concepts and qualitative reasoning:
  1. *Tutorials in Introductory Physics (McDermott et al.)*—A collection of worksheets by the Physics Education Group at the University of Washington (UW).
  2. *ABP Tutorials (Redish et al.)*—Additional tutorials in the UW mode but ones that integrate the technological tools of the Suite—CADAA, extraction of data from videos, and simulations. These also extend the range of topics to include modern physics.

- *Laboratories:* A set of laboratories using CADAA to help students build concepts, learn representation translation, and develop an understanding of the empirical base of physics knowledge. Two levels of labs belong to the Suite.
  1. *RealTime Physics (Thornton, Laws, and Sokoloff):* Appropriate for college level physics.
  2. *Tools for Scientific Thinking (Thornton and Sokoloff):* Similar to RealTime but at a slower pace for high school physics.

The materials of the Suite can be used independently, but their approach, philosophy, and notation are coherent. As a result, you can easily adopt one part as a test of the method when it is convenient and appropriate, or you can integrate several Suite elements, transforming all parts of your class.

Detailed discussions of the various components of the Suite are given in chapters 7–9, and considerations of how they might be used are presented in chapter 10. Those who are familiar with the research and motivation behind modern physics education curriculum reform are invited to turn to those chapters directly. If you are not familiar with the research and theory behind these materials, read the rest of this chapter and the next few chapters where I present some motivation and background.

MOTIVATION

Why do we need the Physics Suite? Most of us learned perfectly well from a text. What is different today? A number of things have changed and are going to be changing even more in the future.

- The students we are teaching have changed.
- The goals we want to achieve with these students have changed.
- We know much more today about how students learn than we used to.
- We have more tools to work with—both technology and new learning environments—than we used to.

I organize my discussion of these points around two questions:

1. Who are we teaching and why?
2. Why Physics Education Research (PER)?

Who are we teaching and why?

Since both the difficulties in teaching physics and their solutions depend on the population of students we are considering, let's begin by considering who our students are—and who they are likely to be in the next few years.

The growth of other sciences

When I began my serious study of science as a high school student more than four decades ago, it seemed to me that only in physics could I do "real" science. By this, I meant discovering fundamental laws of nature and making sense of their implications. As a high school student, I was particularly taken by the beautiful match between the mathematics, which I adored, and the physics, which I took to represent the real world. To a certain extent, I was right—at least for me. Physics is the crown jewel of the sciences, making an elegant link between the understanding of the deeply fundamental and the powerfully practical. Einstein's "$E=mc^2$" and the nuclear bomb that so affected the politics and even the daily sensibilities of many normal citizens during the last half of the last century are only the tip of the iceberg. Quantum mechanics leads us to a deep understanding of the structure of matter, resulting in developments like the transistor and the laser that continue to profoundly change our way of life.

What I missed as a high school student was the immense progress soon to be achieved by the other mature sciences, such as biology and chemistry, and the immense growth soon to be shown in the then infant sciences of computer science and neuroscience, among others. Today, a high school student with an interest in science can have exciting opportunities for a productive and fascinating career in a wide range of sciences, from building models of the universe to modeling neural processes of the brain. Physics is now only one of many mutually enhancing jewels in the crown of science.

Progress in these other sciences has been facilitated by advances in physics in many ways, from improvements in gravitational theories to the development of fMRI (functional magnetic resonance imaging), a tool that uses nuclear magnetic resonance to noninvasively track changes in brain metabolism as people think about different things. Students going into these other sciences need to understand physics as part of their scientific education, but what exactly do they need from us? What role can (and should) physics play as a part of the education of a professional scientist in biology or chemistry? What role can (and should) physics play as a part of the education of a technical professional such as an engineer or paramedic?

The goals of physics for all

Physics instruction has traditionally played two very obvious roles in the education of scientists: both to recruit and train professional physicists-to-be and to "filter out" those students who might not be able to handle the mathematics of engineering or the memorization required in medical school. The former role becomes a smaller fraction of our teaching activities as the number of students studying other sciences grows. The latter no longer seems appropriate for present circumstances, when engineers, scientists, and medical professionals have an increasing need to understand both the systems they are working with and the complex tools they are using to probe them.

Improving our teaching of physics is more important today than ever before. First, a larger fraction of the population is graduating from high school and going on to universities than in previous times. More of these students than ever before are either interested in a scientific career or are concerned about finding jobs in an increasingly technological workplace.

Second, especially for those of us in publicly supported institutions, the governments and the populace that employ us are more likely today to hold the educational system (and therefore its teachers and administrators) directly responsible for the students' learning—or lack of it—than they were in the past. In other times, individual students were seen to be personally responsible for their learning and less attention was paid to the effectiveness of teaching.[2] Today, workplace demands for more technologically trained personnel require that we do whatever we can to help facilitate the successful education of our students.

The task of the physics teacher today is to figure out how to help a much larger fraction of the population understand how the world works, how to think logically, and how to evaluate science. This is doubly important in democratic countries where a significant fraction of the adult population is involved in selecting its leaders—leaders who will make decisions not only on the support of basic science, but on many issues that depend intimately on technological information. It would be of considerable value to have a large fraction of the populace who could not be fooled by the misuse of science or by scientific charlatanism.

Are we already achieving these goals?

Does traditional physics teaching "work" in the introductory physics classroom? Unfortunately, the answer seems to be a resounding "no." Detailed examinations by many physics education researchers have shown that traditional physics instruction does not work well for a large fraction of our students. Many of our students dislike physics; many feel that it has no relation to their personal lives or to their long-term goals; and many fail to gain the skills that permit them to go on to success in advanced science courses.

The nature of the difficulty appears to be a kind of "impedance mismatch." The professor sends out information and sees it reflected back in a similar or identical form (Figure 1.3), but little understanding has actually gotten through to the other side.

Figuring out what doesn't work and what we can do about it

If we are to improve the situation, the best approach is to use our scientific tools to understand what is going on. We need to observe the phenomena we want to understand and try to make coherent sense of what we see. From educators and cognitive psychologists, we learn two important lessons.

- To understand what will work, we have to concentrate on what the student is learning instead of on what we are teaching.

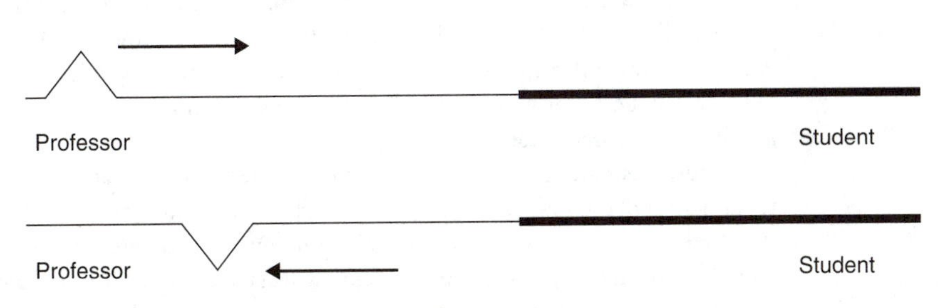

Figure 1.3 The fact that something "comes back as we sent it out" does not mean that much has "gotten through to the student," especially if students possess a large inertia!

[2]In the end, every student is indeed responsible for his or her own learning. But the issue is whether students are to learn everything on their own—no matter what we throw at them—or whether they can learn more with the aid of appropriately designed learning environments and interactions with trained mentors. This is discussed in more detail in chapter 2 under the heading, "The Social Learning Principle."

- We have to do more than evaluate our students' success. We have to listen and analyze what they are thinking and how they learn.

If we really want to change how our students think about the physical world, we have to understand how they think.

Introducing Sagredo

At this point in our discussion, I need to introduce a voice other than my own. Not everything in this book will be obvious to the professional physicist teaching physics, even to one with years of experience. Much of what has been learned in PER is surprising and counterintuitive. Occasionally, contradictory ideas about teaching seem obviously true. In order to make it easier for you to keep track of both sides of the discussions, I introduce my virtual colleague, Sagredo.

Sagredo is a successful researcher at a large research-oriented state university. He is dissatisfied with what his students learn in his introductory physics classes and has not been able to improve the situation despite significant and time-consuming efforts. In a grand old physics tradition [Galileo 1967], I will use Sagredo as a personification of the thoughtful, intelligent physicist who has little or no experience with physics education research. I've chosen Galileo's impartial but intelligent listener, Sagredo, since this book is intended for an audience of professional physicists and physics teachers, most of whom are highly sophisticated in their knowledge of physics but who may not have thought deeply about the issues of how people *think about* physics.

To the last sentence of the previous paragraph, Sagredo might well respond, "I learned physics with traditional instruction from teachers who didn't think about how I thought. Why can't my students do it the same way as I did?" The reason is that we are not only concerned about training physicists.

I recall well the first time I ever taught electromagnetism (to a class of sophomore physics majors at Maryland using Purcell's lovely and insightful text [Purcell 1984]). Suddenly, it seemed, everything made coherent physical sense. Before that, I knew all the equations and could even solve lots and lots of Jackson problems [Jackson 1998] with some alacrity, but I hadn't really "made physics of it" as a graduate student—and I hadn't realized that I hadn't. Amused by the different feeling associated with my new knowledge, I realized that I had studied electromagnetism with Maxwell's equations five times through my high school, college, and graduate school years.

"But," responds Sagredo, "perhaps you needed that. Perhaps one cannot expect someone to understand physics the first time through." Certainly one rarely learns something at the first look. We often have to see something many times before we really learn it. But then we have a problem. Very few of our students will have the opportunity to do what I did—study the physics many times at many levels and eventually teach it. We have to decide the value of one step in a six-step process. If you know your children will take music lessons for 10 years, it might suffice to begin with a year of scales and finger exercises, to strengthen their hands. But it might not. Many, perhaps most, children will rebel and not even make it to a second year.

So the hard question for us is: Is it possible to provide some useful understanding of physics to students in a one-year physics course? Or does the real utility of the course only come in providing a foundation for future physics courses? If the latter is true, few of the students who are now taking introductory high school or university physics would be well advised to continue their efforts, since most will not take any future physics courses. Fortunately, as we begin to understand how students learn physics, we begin to see that remarkable improvements in understanding are possible for a large fraction of students in our introductory courses.

At this point, Sagredo complains: "But if you modify introductory physics so that the average student does better, aren't you going to bore the very best students? These students are important to us! They are the future physicists who will be our graduate students and our successors." Sagredo, I agree that we would need to be concerned if that were the case. If we were to improve our instruction to the middle 50% while degrading it for the top 5%, it could be a disaster for the profession. What is particularly gratifying is that the improved learning that takes place as a result of instructional reform based on an understanding of how students think is not limited to the "middle of the road" student who was previously getting by but was not getting much of long-lasting value from the course. Over the past decade, physics education research has consistently documented that the top students in the class show even stronger gains than the midrange students from research-motivated cognitive-based curriculum reforms. (See, for example, [Cummings 1999].)

WHY PHYSICS EDUCATION RESEARCH?[3]

The ABP Group brings together individual physics education researchers and curriculum developers who are working as a part of a community effort to improve both our understanding and our implementation of physics teaching. An important component of this effort is the word "community." We share the philosophy with a growing cadre of physicists that to teach physics more effectively, we need to work together as a research and development community. We need to work together in the way that scientists in a science and industry sector work together to improve both our knowledge of how the world works and of how to make use of that knowledge. We share the conviction that by using the tools of science—observation, analysis, and synthesis—we can better understand how students learn and find ways to improve how we teach as a result.

Why do we need to go beyond the usual observations we have made, both of our teachers when we were students and of our students now that we are the teachers? To answer this question, we need to think about the nature of our knowledge—both of science and of teaching.

To understand how people learn science and how we might use science to learn about how people learn, we need to think a bit about the nature of the knowledge we are learning. We often say that the goal of science is to discover the laws of nature. This is not quite precise enough for our purposes. It's better to say that we are a community working together to create the best way of thinking about the world that we can. This places the knowledge firmly where it really resides—in the head of the scientist as a part of a scientific community.

[3]Much of this section is based on my Millikan lecture [Redish 1999].

Knowledge as a community map

A good metaphor for the process of science is the building of a map. A map of the world should not be mistaken for the world,[4] but it can nonetheless be of great value in getting around. What is perhaps most important about the scientific map of the world is that it is more than just the collection of the maps of individual scientists. The culture of science includes the continual interaction, exchange, evaluation, and criticism we make of each others' views. This produces a kind of emergent phenomenon I refer to as a *community consensus knowledge base*, or more briefly, a *community map*. I visualize this as an idealized atlas of science. Just as an atlas contains many individual charts, so the atlas of science contains many distinct, coherent, but incomplete areas of knowledge. These areas are supposed to agree where they overlap, but it is not clear that the entire universe can be encompassed in a single map.[5] No single individual, no matter how brilliant, has a map identical to this community consensus map.[6]

At this point, Sagredo might again complain. "But science isn't just the collection of individual scientists' knowledge. What we know in physics is the *correct* description of the real world." Sagredo, I agree. But we have to be a bit more explicit about what this "correct" knowledge is and where it resides. If no one individual has the complete map, but all knowledge ultimately lies in someone's head, in what sense does the knowledge of the world we have gained as a community exist? The key is in the phrase "as a community."

Real maps are constructed in a manner similar to the way we construct science. They are built by many surveyors. No one surveyor has made all the measurements that lead to a map of the United States, for example. Furthermore, each atlas differs in some detail from every other atlas, yet we have little doubt that a true atlas could exist (though it would, of course, have to be dynamic and limited to a preset resolution).

In mathematics, if we have a series of functions that get closer and closer to each other in a prescribed way, then we say the sequence has the Cauchy property.[7] Even if we can't find the true limit analytically, we find it convenient to act as if such a limit exists.[8] The natural mathematical structures of sets of functions behave much more nicely if we "complete" the space by adding the sets of Cauchy sequences to our space. It's like adding the real numbers that fall in between the rationals. We can never calculate them exactly, but it would be very hard to describe the phenomenon of motion if we left them out. (See Figure 1.4.)

In many areas of physics, the sequence of knowledge functions has converged—for all practical purposes. The community consensus on such items as the classical mechanics of the planets of the solar system or the thermodynamics of weakly interacting gases, for example,

[4] Lewis Carroll describes a community of mapmakers who are creating increasingly accurate maps. Finally, they create a map of the area that is 1-1 in scale. Unfortunately, the local inhabitants refuse to let them unroll it "because it will block the sun and kill the crops" [Carroll 1976, p. 265].

[5] Mathematically, this is even true of a sphere, which cannot be mapped by a single non-singular map to a Euclidean plane. See, for example, [Flanders 1963].

[6] In some areas, a specific individual's map may be better than the community's map.

[7] Mathematically stated, a sequence of functions $\{f_n(x)\}$ is said to be *Cauchy* if two functions taken from far enough out in the sequence will be as close together everywhere as you want. (Given any $\epsilon > 0$ there is an N such that if $m,n > N$, $|f_n(x) - f_m(x)| < \epsilon$ for all x.)

[8] For mathematical details, see for example, [Reed 1980, p. 7].

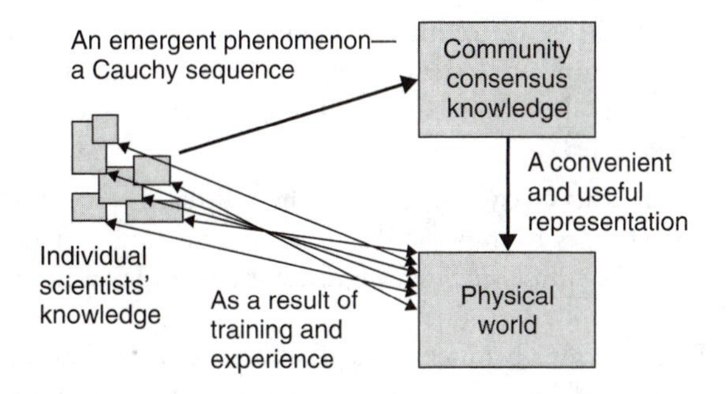

Figure 1.4 Representation of the process of building the scientific map of the physical world.

is exceedingly strong—in part because we know the resolution that is relevant to most problems in these subjects. Just as we don't find it useful to have a map of New York that specifies the cracks in the sidewalk, we don't need to calculate the location of a satellite to nanometer accuracy.

Building the community map for education

If what we learn about physics education is to lead to a stable and growing community map, the community needs to document what we know and to present conjectures and hypotheses for criticisms and questioning. This is particularly important in education.

Human behavior in all realms is beset by wishful thinking—we tend to really believe that what we *want* to be true *is* true. To some extent, the most important part of the process by which science builds its community consensus knowledge base is the part that identifies and purges the wishful thinking of individual scientists. Some parts of the process critical for this task include:

- *Publication* of results, documented with sufficient care and completeness that others can evaluate and duplicate them
- *Repetition* of experiments using different apparatus and different contexts[9]
- *Evaluation* and critiquing of one scientist's results by others through refereeing, presentations and discussions in conferences, and through follow-up evaluations and extensions

When it comes to education, wishful thinking is not just present; it is widespread and can take a variety of forms.

- A dedicated and charismatic teacher may, by force of personality, inspire her students into learning far above the norm. That teacher may then try to disseminate her curriculum to other less charismatic individuals, only to find the method is no longer effective.

[9]We try to make experiments as similar as possible, but it is not, of course, possible to ever reproduce an experiment exactly—even if the identical apparatus is used. These small variations help us understand what variables are important (e.g., the colored stripes on the resistors) and which are not (e.g., the color of the insulation on the wires).

- A teacher delivering an inappropriately rigorous course may find that his students seem to learn little and to dislike it intensely. "Ah," he is heard to remark, "but when they're older they will realize that I was right and come to appreciate the course and what they've learned."
- A teacher concerned about how little his students are learning may try a number of changes to improve the situation, but find that nothing seems to help. "Oh well," he says, "those students are just not able to learn physics under any circumstances."[10]

I have personally observed each of these responses from physics colleagues whose science and whose teaching efforts I respect. Each of these situations is more complex than the individual teacher has realized. In each of these situations, much more can be done if a deeper understanding of learning and teaching is brought to bear.

Building a community knowledge base about education requires using our full array of scientific tools—observation, analysis, synthesis, plus the community purging and cleaning tools of publication, repetition, and evaluation. Sagredo complains at this point. "Do we really need all this effort? Oh, I know you're right about some teachers, and I recognize your three 'wishful thinkings'—I've occasionally been there myself. But there are some good teachers. I've had one or two. Why don't we just let them concentrate on the teaching and carry most of the introductory load?"

Yes, Sagredo, there have always been excellent teachers—teachers who can reach not only the very best students, but who can energize and educate even the less motivated and less capable students. But there are far too few of them, and their skill has not been easily transferable to other concerned and dedicated, but less naturally talented, teachers. We want to understand what it is those teachers are doing successfully so as to be able to transform successful teaching from an art, practiced only by a few unusual experts, to a technology that can be taught, learned, and facilitated by powerful tools. Building an understanding of the educational process through using the tools of science is beginning to enable us to carry out this transformation.

The impact on teaching of research on teaching and learning

Sagredo is still not convinced. "How can education research in someone else's class tell me anything about what is happening in my own? Every situation is different—different students, different teacher, different university. Each of those differences is important." You are right, Sagredo, there are differences and they matter. But a lot of what has been learned is robust enough to illuminate what is happening in many different situations. Let me illustrate this with specific examples of how research has affected two instructors.

Even good students get the physics blues

An example of how building a scientific community of physics education researchers can spread and transform individuals is told by Eric Mazur, a chaired professor at Harvard

[10]Note from this example that wishful thinking does not necessarily imply a rosy view of a situation. It may be that the wishful thinking is that "the situation is so bad that there is nothing *I* can do about it and therefore I don't have to make an effort."

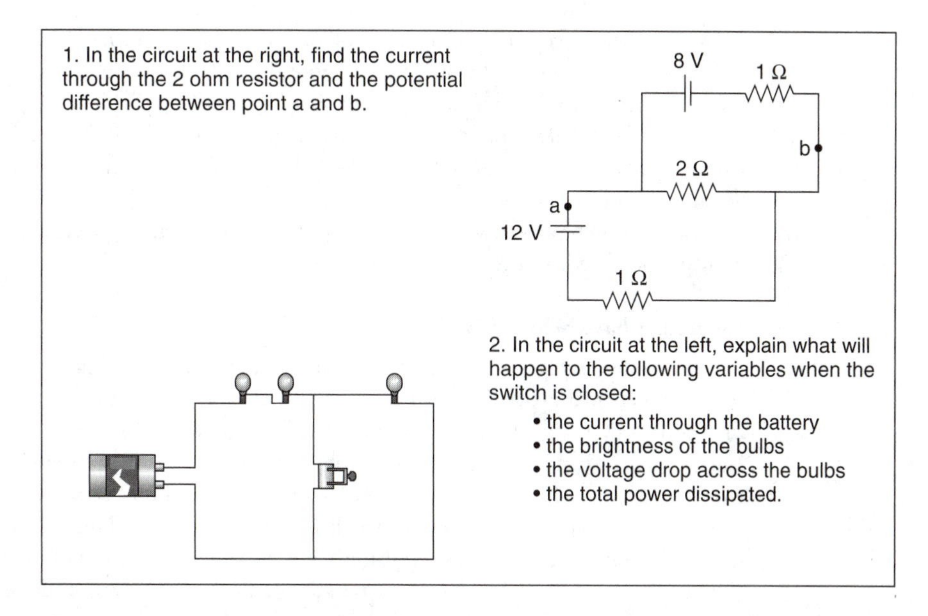

Figure 1.5 A quantitative and a qualitative problem in DC circuits (from [Mazur 1997]).

University. Mazur read the paper published by Ibrahim Halloun and David Hestenes in 1985 in which they described common student conceptual difficulties revealed by physics education research [Halloun 1985a]. Mazur was quite skeptical, being reasonably satisfied with the grades his students achieved on his examinations. Halloun and Hestenes had included a survey instrument in their paper, a 29-item multiple-choice test probing students' understanding of fundamental concepts in mechanics.[11] Mazur looked at the questions and found them trivial. He was certain his Harvard students would have no trouble with any of the questions. He decided to give the test to his introductory physics class after appropriate instruction [Mazur 1992]. Upon looking at the questions, one student asked: "Professor Mazur, how should I answer these questions? According to what you taught us, or by the way I *think* about these things?"[12] Mazur was appalled at how many "trivial" questions his students missed. (See, for example, the problem shown in Figure 4.1.) He began to look at his teaching as a research problem.

Mazur went on to study and document in detail the difference between the algorithmic problem-solving skills his students displayed and the conceptual understanding that he had been assuming automatically came along for the ride [Mazur 1997]. On an examination to his algebra-based physics class, he gave the problems shown in Figure 1.5.

[11] An updated version of this exam, the Force Concept Inventory (FCI), is on the Resource CD distributed with this volume. Also see the discussion of the FCI in chapter 5.

[12] This confirms both Mazur's and my prejudice that Harvard students are good students. Many students have this same dichotomy but are not aware that it exists. The Harvard students actually did well on the exam—given our current understanding of what can be expected on this exam after traditional instruction—just not as well as Mazur expected.

The average score on the first problem was 75%; the average score on the second was 40%. Students found problem 2 much more difficult than problem 1, despite the fact that most physicists would consider the analysis of the second problem, the short circuit, much simpler; indeed, parts of it might be considered trivial. This study and many others[13] show that students frequently can solve complex algorithmic problems without having a good understanding of the physics.

Before his epiphany, Mazur was a popular and entertaining lecturer. After his encounter with physics education research, he became a superb and effective *teacher*.

I wouldn't have believed it if I hadn't seen it

I can provide a second example of the value of the community exchange of research results from my personal experience. When Arnold Arons' book on teaching physics [Arons 1990] first appeared, I was absolutely delighted. Although I was not yet a physics education researcher, I had had a strong interest in physics teaching for many years. I had read many of Arons' papers and had great respect for them. I read the book cover to cover and annotated it heavily. In chapter 6 (p. 152) you will find the sentence: "This paves the way for eliminating misconceptions such as repulsion between a north magnet pole and a positive electric charge, and so on." I wasn't very worried about this. It isn't even underlined in my copy of Arons. (I underlined about a fifth of the sentences in that chapter.)

But in January of 1994, the Physics Education Group (PEG) at the University of Washington reported the results of a study of engineering students' responses to being taught about magnets [Krause 1995]. Traditionally, many teachers and textbook writers assume, just as I did, that students know little about the subject, so a good way to introduce it is by analogy with electric charge, the topic typically presented just before magnetism. The Washington PEG demonstrated that before the lectures on magnetism, more than 80% of their engineering students confused electric charges and magnetic poles as measured by the simple problem shown in Figure 1.6. After traditional instruction, this number remained above 50%. I was both flabbergasted and distressed at hearing this. I had taught the subject off and on for nearly 25 years and was teaching it at the time of the presentation. Furthermore, I believed that I listened carefully to students, and I was already sensitized to the issue that students bring their previous knowledge to any new learning experience. Yet I had never imagined such a confusion was common. I probed my class upon my return and, needless to say, found exactly the same results as the Washington group.

The Arons book is still one of the best "teacher-to-teacher" books available. But despite my respect for Arons' insights, I was skeptical about the importance of a possible student confusion between electric charge and magnetic poles. Indeed, I felt my personal experience contradicted it. The point was only convincingly brought home to me by the solid experimental data offered by the UWPEG.[14]

[13]For example, [Halloun 1985b].

[14]Note further that this result had been known previously and even published, but not in a journal that I looked at regularly or that was conveniently available. See [Maloney 1985].

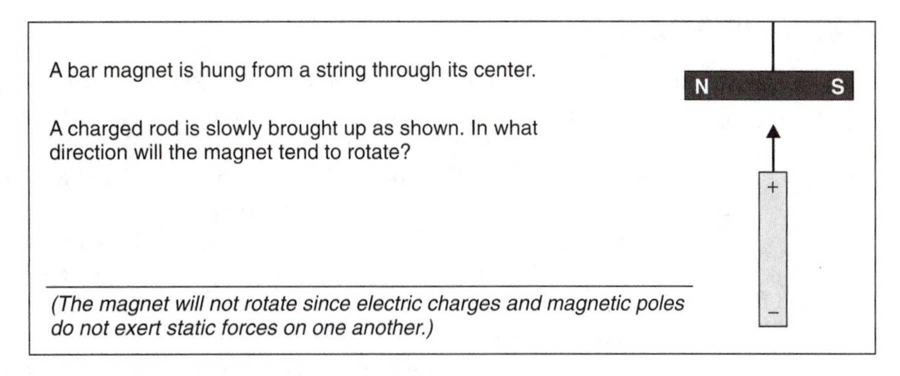

Figure 1.6 Problem that reveals student confusions about electric and magnetic poles.

SOME CAVEATS

Education research deals with an extremely complex system. At present, neither the educational phenomenology growing out of observations of student behavior nor the cognitive science growing out of observations of individual responses in highly controlled (and sometimes contrived) experiments has led to a single consistent theoretical framework. Indeed, it is sometimes hard to know what to infer from some particular detailed experimental results.

Yet those of us in physics know well that advancement in science is a continual dance between the partners of theory and experiment, first one leading, then the other. It is not sufficient to collect data into a "wizard's book" of everything that happens. That's not science. Neither is it science to spout high-blown theories untainted by "reality checks." Science must build a clear and coherent picture of what is happening at the same time as it continually confirms and calibrates that picture against the real world.

At present, the combination of education research, cognitive research, and neuroscience does not provide us with a consistent or coherent picture of how students' minds function when they learn physics. Indeed, many problems have been caused by inappropriately generalizing observations or by interpreting some tentative psychological theory as hard and fast. Using rules generated by behavioral research incautiously without reference to continual checks against experimental data (in the classroom!) can lead us to the wrong conclusions.

But in many cases, educational research has been able to tell us what does *not* work. And although it does not provide prescriptive solutions, I have found that the results of educational research and cognitive science help me to organize my thinking about my students and to refocus my attention. Instead of concentrating only on the organization of the physics content, I now also pay attention to what my students are doing when they interact with a physics course. This is not to suggest that an emphasis on content is somehow unimportant or should be neglected. What we are teaching is important, but it must be viewed in the context of how our students learn.

WHAT THIS BOOK IS ABOUT

In this book, my goal is to provide a guide for teachers of physics who are interested in implementing some of the best modern methods that have been developed as a result of the community's taking a scientific approach to figuring out how to teach physics. The elements of the Physics Suite (as well as some others that match well and are easily integrated into a course using Suite elements) are discussed in chapters 6–10.

It is important to realize, however, that although excellent student-centered approaches to teaching physics have been developed, none of them are "plug-and-play." Student-centered instruction doesn't mean students are left on their own to do whatever they choose. These modern approaches require that instructors provide their students with substantial guidance and learn to work with their students in new ways. That requires that the instructor be reasonably well informed about the premises and methods that are being used.

Most of the literature that backs up these new methods discusses student difficulties together with explicit data documenting the frequencies and environments in which those difficulties occur, when they are known. (See the Resource Letter on Physics Education Research on the Resource CD distributed with this volume [McDermott 1999].)

As a theoretical physicist, I am uncomfortable with providing masses of data without trying to put them into a theoretical frame. Having such a frame helps make sense both of what is seen in the research literature and what is seen in the classroom. The appropriate frame for making sense of educational data is an understanding of how students (and people in general) think and reason. Therefore, I include a fairly extensive chapter on the relevant elements of cognitive science and their implication for instruction (chapter 2). This may seem strange as a component of what is essentially a physics book, but teaching physics is not only about physics: it is also about how we *think* about physics.

So it would be useful to have some understanding of how our students think. Sagredo is nervous about this. "I studied psychology in college. It was all about silly, unrepeatable things like dreams or irrelevant things like rats running mazes. There were all kinds of conflicting schools and fads that came and went. Is there anything really useful there for us?" Yes, Sagredo, I too took psychology in college (in 1960) and came away mostly disappointed, but much has happened since then. There are still schools and conflicting opinions, but interestingly enough, there is beginning to emerge a broad consensus, at least on a number of elements that can be useful for us.

In the next two chapters, I discuss those elements of a cognitive model of thinking and learning that are relevant for physics education, and I give guidelines and heuristics that can help us better understand and improve our teaching.

CHAPTER 2

Cognitive Principles and Guidelines for Instruction[1]

He who loves practice without theory
is like the sailor who boards ship
without a rudder and compass
and never knows where he may cast.
Leonardo da Vinci
quoted in [Fripp 2000]

When we present instruction to our students, we always build in our assumptions: our expectations as to what students will do with whatever we give them, our assumptions about the nature of learning, and our assumptions about the goals of our particular instruction. Sometimes those assumptions are explicit, but more frequently they are unstated and rarely discussed. Some pertain to choices we get to make, such as the goals of instruction. Others are assumptions about the nature and response of a physical system—the student—and these are places where we can be right or wrong about how the system works.

If we design our instruction on the basis of incorrect assumptions about our students, we can get results that differ dramatically from what we expect. To design effective instruction—indeed to help us understand what effective instruction means—we need to understand a bit about how the student mind functions. Much has been learned about how the mind works from studies in cognitive science, neuroscience, and education over the past 50 years. In this chapter and the next, I summarize the critical points of the cognitive model and organize the information in a way that relates to the instructional context. I then consider some specific implications for physics instruction: the impact of considering students' prior knowledge and the relevant components of physics learning other than content. The chapter ends with a discussion of how our explicit cognitive model of student learning can provide guidelines to help us both understand what is happening in our classroom and improve our instruction.

[1]This chapter is based in part on the paper [Redish 1994].

THE COGNITIVE MODEL

To understand learning, we must understand memory—how information is stored in the brain. Modern cognitive science now has complex and detailed structural information about how memory works. For some simple organisms like the marine snail *Aplysia*,[2] the process is understood down to the level of neuron chemistry [Squire 1999]. We don't need that level of detail for the "application" of understanding physics teaching and learning. A few simple principles will help us understand the critical issues.

Models of memory

It is clear, from all the different things that people can do that require memory, that memory is a highly complex and structured phenomenon. Fortunately, we only need to understand a small part of the structure to get started in learning more about how to teach physics effectively. There are a few critical ideas that are relevant for us. First, memory can be divided into two primary components: *working memory* and *long-term memory*.

- Working memory is fast but limited. It can only handle a small number of data blocks, and the content tends to fade after a few seconds.
- Long-term memory can hold a huge amount of information—facts, data, and rules for how to use and process them—and the information can be maintained for long periods (for years or even decades).
- Most information in long-term memory is not immediately accessible. Using information from long-term memory requires that it be activated (brought into working memory).
- Activation of information in long-term memory is productive (created on the spot from small, stable parts) and associative (activating an element leads to activation of other elements).

In the rest of this section, I elaborate on and discuss some of the characteristics of memory that are particularly relevant for us.

1. Working memory

Working memory appears to be the part of our memory that we use for problem solving, processing information, and maintaining information in our consciousness. Cognitive and neuroscientists have studied working memory rather extensively. Not only is it very important to understand working memory in order to understand thinking, but working memory can be studied with direct, carefully controlled experiments [Baddeley 1998]. For our concerns here, two characteristics are particularly important:

- Working memory is limited.
- Working memory contains distinct verbal and visual parts.

[2] *Aplysia* has a nervous system with a very small number of neurons—about 20,000, some of them very large—and a very simple behavioral repertoire. As a consequence, it is a favorite subject for reductionist neuroscientists. See, for example, [Squire 1999].

Working memory is limited. The first critical point about working memory for us to consider is that working memory can only handle a fairly small number of "units" or "chunks" at one time. Early experiments [Miller 1956] suggested that the number was "7 ± 2". We cannot understand that number until we ask, "What do they mean by a unit?" Miller's experiments involved strings of numbers, letters, or words. But clearly people can construct very large arguments! If I had to write out everything that is contained in the proof of a theorem in string theory, it would take hundreds, perhaps thousands, of pages. The key, of course, is that we don't (write out everything, that is). Our knowledge is combined into hierarchies of blocks (or chunks) that we can work with even with our limited short-term processing ability.

You can see the structure of working memory in your own head by trying to memorize the following string of numbers:

3 5 2 9 7 4 3 1 0 4 8 5

Look at it, read it aloud to yourself, or have someone read it aloud to you; look away for 10 seconds and try to write the string down without looking at it. How did you do? Most people given this task will get some right at the beginning, some right at the end, and do very badly in the middle. Now try the same task with the following string

1 7 7 6 1 8 6 5 1 9 4 1

If you are an American *and* if you noticed the pattern (try grouping the numbers in blocks of four), you are likely to have no trouble getting them all correct—even a week later.

The groups of four numbers in the second string are "chunks"—each string of four numbers is associated with a year, and is not seen as four independent numbers. The interesting thing to note here is that some people look at the second string of numbers and do not automatically notice that it conveniently groups into dates. These people have just as much trouble with the second string as with the first—until the chunking is pointed out to them. This points out a number of interesting issues about working memory.[3]

- Working memory has a limited size, but it can work with chunks that can have considerable structure.
- Working memory does not function independently of long-term memory. The interpretation and understanding of items in working memory depend on their presence and associations in long-term memory.
- The effective number of chunks a piece of information takes up in working memory depends on the individual's knowledge and mental state (i.e., whether the knowledge has been activated).

This second point is fairly obvious when we think about reading. We see text in terms of words, not in terms of letters, and the meanings of those words must be in long-term

[3]Of course in this example, it is not simply the chunking that makes the numbers easy to recall. It is the strong linkage with other, semantic knowledge.

storage. The third point is something we will encounter again and again in different contexts: How students respond to a piece of information presented to them depends both on what they know already and on their mental state—what information they are cued to access.

The number of chunks a piece of information has for an individual depends not only on whether or not they have relevant associations but on how strong and easily accessible that knowledge is in long-term memory. When a bit of knowledge—a fact or process—is easily available and can easily be used as a single unit in working memory, we say the knowledge is *compiled.* Computer programming is a reasonably good metaphor for this. When code in a high-level computer language has to be translated line by line into machine instructions, the code runs slowly. If the code is compiled directly so that only machine-language instructions are presented, the code runs much more quickly.

Some of the difficulties students encounter—and that we encounter in understanding their difficulties—arise from this situation. Physics instructors work with many large blocks of compiled knowledge. As a result, many arguments that seem simple to them go beyond the bounds of working memory for their students. If the students have not compiled the knowledge, an argument that the instructor can do in a few operations in working memory may require the student to carry out a long series of manipulations, putting some intermediate information out to temporary storage in order to carry out other parts of the reasoning.

Studies with subjects trying to recall strings of information indicate that items fade from working memory in a few seconds if the subject does not try to remember the information by repeating it consciously [Ellis 1993]. This working memory repetition is known as *rehearsal.* Think about looking up a telephone number in a phonebook. Most of us can't remember it—even for the few seconds needed to tap in the number—without actively repeating it.

The short lifetime of working memory has serious implications for the way we communicate with other people, both in speaking and in writing. In computer science, the holding of information aside in preparation for using it later is called *buffering,* and the storage space in which the information is placed is called a *buffer.* Since human memory buffers are volatile and have a lifetime of only a few seconds, it can be very confusing to present information that relies on information that has not yet been provided. Doing this can interfere with a student's ability to make sense out of a lecture or with a websurfer's ability to understand a webpage.[4]

Working memory contains distinct verbal and visual parts. A second characteristic of working memory that has been well documented in the cognitive literature[5] is that working memory contains distinct components. At least the verbal component (the *phonological loop*) and the visual component (the *visual sketchpad*) of working memory appear to be distinct. (I am not aware of any evidence in the cognitive literature about the independence of other components such as mathematical or musical.) This has been demonstrated by showing that two verbal tasks or two spatial tasks interfere with each other substantially more than a visual task interferes with a verbal task, or vice versa.[6]

[4] In the theory of communications, this leads to the "given-new principle" in conversation [Clark 1975] and writing [RedishJ 1993].

[5] See [Baddeley 1998] and [Smith 1999] and the references therein.

[6] The evidence for this is also strong from neurophysiological lesion studies [Shallice 1988].

2. Long-term memory

Long-term memory is involved in essentially all of our cognitive experiences—everything from recognizing familiar objects to making sense of what we read. An important result is that recall from long-term memory is *productive* and *context dependent.*

Long-term memory is productive. What we mean here by "productive" is that memory response is active. Information is brought out of long-term storage into working memory and processed. In most cases, the response is not to simply find a match to an existing bit of data, but to build a response by using stored information in new and productive ways. This construction is an active, but in most cases, an automatic and unconscious process. Think of language learning by a small child as a prototypical example. Children create their own grammars from what they hear.[7] Another model of the recall process is computer code. A result, such as sin(0.23 rad), may be stored as tables of data from which one can interpolate or as strings of code that upon execution will produce the appropriate data. Analogs of both methods appear to be used in the brain.

Another example demonstrates that it's not just sensory data that the brain is processing; cognitive processes such as recall and identification of objects is also productive. Look at the picture in Figure 2.1. It consists of a number of black blobs on a white background. The subject of the picture will be immediately obvious to some, more difficult to see for others. (See the footnote if you have looked at the picture for a while and can't make it out.[8]) Even though you may never have seen the particular photograph from which this picture was constructed, your mind creates recognition by pulling together the loosely related spots and "constructing" the image. Once you have seen it, it will be hard to remember what it looked like

Figure 2.1 A picture of an animal [Frisby 1980].

[7] The fact that they don't always create the same rules as their parents have is one of the facts that causes languages to evolve.

[8] The picture is of a Dalmatian dog with head down and to the left, drinking from a puddle in the road, seen through the shadows under a leafy tree.

to you when you couldn't see it. When you couldn't "see" the picture in the blobs, was there a picture there? Now that you see it, where is the picture, on the paper or in your brain?

Long-term memory is context dependent. By *context-dependent,* I mean that the cognitive response to a mental stimulus depends on both (1) the external situation and the way in which the stimulus is presented and (2) the state of the respondent's mind when the stimulus is presented. The first point means, for example, that for a problem on projectile motion presented to a student in a physics class, a student might bring out of long-term memory a different repertoire of tools than the ones she might access if the same problem arose on the softball field. To see what the second point means, consider for example, a situation in which a student is asked to solve a physics problem that could be solved using either energy or force methods. If the problem is preceded by a question about forces, the student is much more likely to respond to the problem using forces than if the question were not asked [Sabella 1999].

To show how the context dependence affects the resources that one brings to the analysis of a situation, let's look at the following example.[9] Suppose I am holding a deck of 3 × 5 file cards. Each card has a letter on one side and a number on the other. I remove four cards from the deck and lay them on the table as shown in Figure 2.2.

I make the following claim: *This set of four cards satisfies the property that if there is a vowel on one side of the card, then there is an odd number on the other.* How many cards do you need to turn over to be absolutely certain that the cards have been correctly chosen to satisfy this property?

Try to solve this problem before looking to the footnote for the answer.[10] Careful! You have to read both the claim and the question carefully. A similar problem but with a different context is the following.

You are serving as the chaperone and bouncer[11] at a local student bar and coffee house. Rather than standing at the door checking IDs all the time, you have occupied a table so you can do some work. When patrons come in and give their order, the servers bring you cards

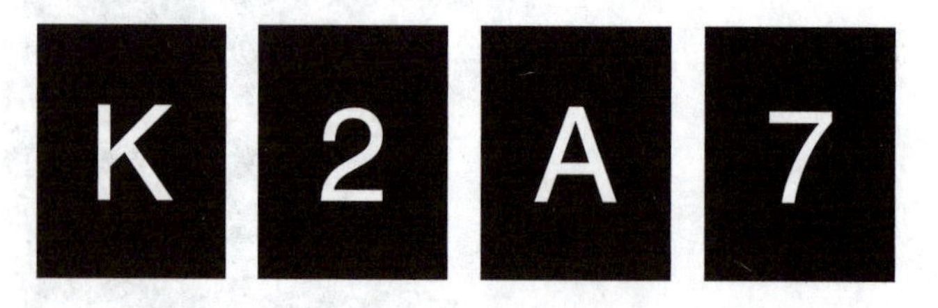

Figure 2.2 An abstract problem. (See text.)

[9]Adapted from [Wason 1966] and [Dennett 1995].

[10]You may have to turn over at most two cards to be sure I am telling the truth. (If the first card fails, you know I am wrong right away.) The only cards that are relevant are the number "2" and the letter "A." Note that the statement only says "if," not "if and only if." To test whether $p \rightarrow q$, you have to test the equivalent statements: $p \rightarrow q$ and $\sim q \rightarrow \sim p$.

[11]This is a highly culture-dependent example. In order to solve it, you must know that in most American communities, the law prohibits the purchase of alcoholic beverages by individuals younger than 21 years of age. Putting the problem into this cultural context also broadens the concerns of some respondents as well as their tools for solving it. Some worry whether or not the servers can be trusted or might be lying if, for example, a friend were involved.

Figure 2.3 A more concrete card problem. (See text.)

with the patron's order on one side and their best guess of the patron's age on the other. You then decide whether to go and check IDs. (The servers can be assumed to be trustworthy and are pretty good guessers.)

A server drops four cards on the table. They land as shown in Figure 2.3. Which cards would you turn over in order to decide whether to go back to the table to check IDs?[12]

This problem is mathematically isomorphic to the previous one, yet most American adults find the first problem extremely difficult. They reason incorrectly or read the problem in an alternative way that is plausible if a word or two is ignored or misinterpreted. I have presented these problems in lectures to physics departments many times. More than two-thirds of physicists produce the wrong answer to the K2A7 problem after a minute or two of consideration. Almost everyone is able to solve the second problem instantly.

These problems provide a very nice example of both productive reasoning and context dependence. In the two cases, most people call on different kinds of reasoning to answer the two problems. The second relies on matching with social experience—a kind of knowledge handled in a much different way from mathematical reasoning.

This result has powerful implications for our attempts to instruct untrained students in physics. First, it demonstrates that the assumption that "once a student has learned something, they'll have it" is not correct. Most physics instructors who have tried to use results the students are supposed to have learned in a math class are aware of this problem. The example shows that even changing the context of a problem may make it much more difficult. Second, it points out that a problem or reasoning that has become sufficiently familiar to *us* to feel like 16/Coke/52/Gin & tonic may feel like K2A7 to our students! We need to maintain our patience and sympathy when our students can't see a line of reasoning that appears trivial to us.

Long-term memory is structured and associative. The examples in the previous subsections illustrate the fundamental principle that the activation of knowledge in long-term memory is structured and associative. When a stimulus is presented, a variety of elements of knowledge may be activated (made accessible to working memory). The particular elements that are activated can depend on the way the stimulus is presented and on the state of the mental system at the time (the context). Each activation may lead to another in a chain of *spreading activation* [Anderson 1999].

[12] You would only have to turn over the cards labeled "16" and "Gin & Tonic." You are not concerned with what a person clearly much older than 21 orders, and anyone is allowed to order a coke.

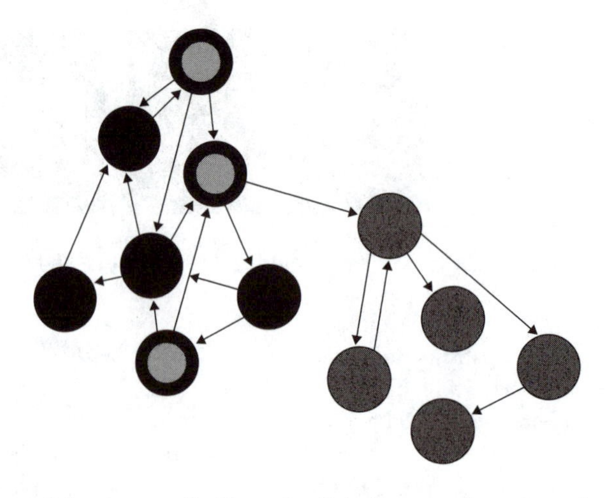

Figure 2.4 An example illustrating linked context-dependent schema.

The key to understanding student reasoning is understanding the *patterns of association* that activate knowledge elements. In general, a pattern of association of knowledge elements is sometimes referred to as a *knowledge structure.* A pattern that tends to activate together with a high probability in a variety of contexts is often referred to as a *schema* (plural, *schemas* or *schemata*) [Bartlett 1932] [Rumelhart 1975]. This is illustrated in Figure 2.4. Each circle represents an element of knowledge. Arrows indicate a probability that the activation of one element will lead to the activation of another. The different colorings of the circles indicate schemas—associated knowledge elements that tend to be activated together. Notice that on the left, some elements have multiple circles. This indicates that particular knowledge elements can activate different schemas, depending on context.

As an example, consider meeting a new person at a beach party. In your conversation with this individual, you activate a number of responses—seeking knowledge of your own about topics the other person raises, looking for body language that demonstrates interest in continuing the conversation, and so on. If, at some point later in the party, the individual falls and is knocked unconscious, a different chain of knowledge and responses is activated. Is the person seriously injured? Do you need to get them to an emergency room? Should the person be moved or medical personnel called? You still begin with a response to the individual as a person, but different associated knowledge elements are activated.[13]

When a schema is robust and reasonably coherent I describe it with the term *mental model.* Since scientific models tend to be organized around the existence, properties, and interaction of objects, when a mental model has this character I refer to it as a *physical model.* A physical model may or may not agree with our current community consensus view of physics. For example, the phlogiston picture of thermodynamics was organized around an

[13] In reading this situation, did you envision the person as an individual as the same or opposite sex to your own? This component of context is an example of context affecting activation responses.

image of a physical substance that we now understand cannot be made consistent with physical observations, so this physical model does not agree with our current community physical model. Two individuals well versed in a topic may have different physical models of the same system. For example, a circuit designer may model electric circuits in terms of resistors, inductors, capacitors, and power sources—macroscopic objects with particular properties and behaviors. A condensed-matter physicist may think of the same system as emerging from a microscopic model of electrons and ions.

Cognitive resources for learning

The fact that the mind works by context-dependent patterns of association suggests that students reason about physics problems using what they think they know by generalizing their personal experience. This doesn't sound surprising at first, but we may be surprised at some of its implications.

When I learned[14] that students in introductory physics often bring in a schema of motion that says objects tend to stop, I enthusiastically presented the result to Sagredo. He was skeptical and argued "If you don't tell them about friction, they wont know about it." Sorry, Sagredo. They may not know the word "friction" (they mostly do) or the rules of physics that describe it (they mostly don't), but they are very familiar with the fact that if you push a heavy box across the floor, it stops almost as soon as you stop pushing it. They also know that if they want to move, they have to exert an effort to walk and when they stop making that effort, they stop.

"But," responds Sagredo, "if you push a box very hard along a slippery floor it will keep going for quite a distance. If you run and stop making an effort, you'll continue going and fall over. Surely they know those facts as well." Absolutely right, Sagredo. But the problem is that most students do not attempt to make a single coherent picture that describes all phenomena. Most are satisfied with a fairly general set of often-inconsistent rules about "how things work."

Over the years there have been some disagreements among researchers as to the nature of the schemas students bring with them to the study of introductory physics. Some researchers suggested that students had "alternative theories"—reasonably self-consistent models of the world that were different from the scientist's [McCloskey 1983]. But extensive work by a number of different researchers (see in particular the work of McDermott [McDermott 1991] and diSessa [diSessa 1993]) suggests that student knowledge structures about physics tend to be weak patterns of association that rarely have the character of a strong schema.[15]

Although I concur with this view, I note that occasionally a student's schema may be more coherent than we as scientists tend to give credit for, since we analyze the student's views through the filter of our own schemas. For example, a student who thinks about "motion" and fails to separate velocity from acceleration may seem inconsistent to a physicist, whereas, in fact, the student may feel he is consistent but may have a model that only works for a rather limited set of specific situations and questions. As long as the student is either aware

[14] I learned this from reading the paper of Halloun and Hestenes [Halloun 1985a] that also inspired Mazur—see chapter 1.

[15] By "strong" or "weak" we simply mean a high or low probability of activating a link to other related (and appropriate) items in a student's schema.

of those limitations or not presented with situations in which that model doesn't work, the student can function satisfactorily.

There are two reasons why it is important for us to understand the knowledge and reasoning about the physical world that students bring with them into our classes. First, it helps us understand what errors the students will commonly make and how they will misinterpret what we say and what they read. We can then use this understanding to help us design both new and improved instruction and better evaluations. I find understanding student thinking particularly useful in helping me to answer students' questions, both one-on-one during office hours and during lecture. It is very easy to misinterpret a student's question as being much more sophisticated than it really is. Knowing common lines of reasoning or associations can help probe the individual student as to what his or her problem really is. In lecture it can help us understand what a question that seems a "stupid" question on its face may really be telling us about the state of many of the students in our class and our interaction with them.

Second, the students bring into our class the basic *resources* from which they will build their future knowledge. Since new knowledge is built only by extending and modifying existing schema, students' existing knowledge is the raw material we have to work with to help them build a more correct and more scientific knowledge structure [Hammer 2000] [Elby 1999].

The knowledge and reasoning that students bring into our class have been analyzed in three ways that are useful for us: (1) as common naïve conceptions, (2) in terms of primitive reasoning elements, and (3) in terms of the way reasoning and knowledge are situated in everyday real-life circumstances. The last-named is called *situated cognition.*

1. Robust reasoning structures: Common naïve conceptions

Occasionally, students' patterns of associations concerning physical phenomena are strikingly robust—they occur with a high enough probability in many contexts for us to refer to them as mental models. In many cases, they contain inappropriate generalizations, conflations of distinct concepts (such as treating velocity and acceleration as a single concept, "motion"), or separations of situations that should be treated uniformly (such as treating a box sliding along a rough floor and a rapidly moving baseball using different rules). When a particular mental model or line of reasoning is robust and found in a significant fraction (say on the order of 20% of students or more), I refer to it as a *common naïve conception.* In the education research literature, these patterns are often referred to as *misconceptions, alternative conceptions,* or *preconceptions,* particularly when they lead to incorrect predictions or conclusions. I choose the more descriptive term, "common naïve conceptions," rather than the most common parlance, "misconceptions," both because it lacks the pejorative sting and because I want to emphasize the complexity of the student concept. Usually, these conceptions are not "just wrong." Students may be naïve, but they're not fools. Their naïve conceptions are usually valuable and effective in getting them through their daily lives. Indeed, most naïve conceptions have kernels of truth that can help students build more scientific and productive concepts.

The presence of common naïve conceptions really isn't so surprising if we think about our students' previous experience. Why should we be surprised that students think that any moving object will eventually come to a stop? In their direct personal experience that is always the case. It's even the case in the demonstrations we show in class to demonstrate the opposite! When I slide a dry-ice levitated puck across the lecture table, I catch it and stop it at the end of the table. If I didn't, it would slide off the table, bounce, roll a short distance, and stop. Every student knows that. Yet I ask them to focus on a small piece of the demonstration—the stretch of about four or five seconds when the puck is sliding along the table smoothly—and extend that observation in their minds to infinity. The student and the teacher may focus on different aspects of the same physical phenomena.[16]

Many teachers show surprise when they learn the results of physics education research demonstrating that students regularly generalize their naïve schemas incorrectly. Why should it be surprising that students think cutting off part of a lens will result in only part of an image being visible on the screen [Goldberg 1986]? Try looking through a magnifying glass! (Yes, I know that's not a real image.) Where do students get the idea that electricity is something that is used up in a resistor [Cohen 1983]? We've told them that you need circuits and that currents flow in loops! Although we don't always think about it, most of our students have had extensive experience with electricity by the time they arrive in our classes. When I said the current had to come in one wire and go out the other, one of my students complained: "If all the electricity goes back into the wall, what are we paying for?"

Much of the effort in the published physics education research literature has been to document the common naïve conceptions of introductory physics students and to develop instructional methods to deal with them. To get an introduction to this literature, consult the papers listed in the Resource Letter given on the Resource CD [McDermott 1999]. A good overview is given in the books [Arons 1990], [Viennot 2001], and [Knight 2002].

2. Modular reasoning structures: Primitives and facets

Perhaps the most extensive and detailed analysis of student reasoning in introductory physics has been diSessa's monumental work, "Toward an Epistemology of Physics" [diSessa 1993].[17] In this study, diSessa analyzes the evolution of reasoning in mechanics of 20 MIT students in introductory calculus-based physics. Although this is a fairly small number of students and a rather narrow population, the care and depth of the analysis make it worthy of attention.[18] Subsequent investigations show the presence of diSessa's results in much broader populations. As of this writing, diSessa's approach has only rarely been applied to the development of new curriculum. Nonetheless, because of the powerful insights it provides into student reasoning, I believe it will be of use both in future curriculum reform and in research trying to understand student thinking, and so I include a brief discussion of his ideas here.

DiSessa investigated people's *sense of physical mechanism*, that is, their understanding of "Why do things work the way they do?" What he found was that many students, even after instruction in physics, often come up with simple statements that describe the way they think things function in the real world. They often consider these statements to be "irreducible"—as the obvious or ultimate answer; that is, they can't give a "why" beyond it. "That's just the way things work," is a typical response. DiSessa refers to such statements as *phenomenological primitives.*

Some of these primitives may activate others with a reasonably high priority, but diSessa claims that most students have rather simple schemas. Primitives tend to be linked directly to a physical situation. They are recognized in a physical system rather than derived by a long chain of reasoning.

[16] This argument is made in a slightly different context in [KuhnT 1970].

[17] A shorter, more accessible introduction to diSessa's ideas is given in [diSessa 1983].

[18] Note that the point of this kind of study is to determine the range of possibilities, not their distribution among a particular population. As a result, the small number of students in the study is not a serious drawback.

As an example, consider the response of a student to pushing a box on a rough surface. The student might respond that "you need a big force to get it going" (*overcoming*: one influence may overpower another by increasing its magnitude), but then "you need a force to keep it going" (*continuous push*: a constant effort is needed to maintain motion). DiSessa identifies the parentheticals as primitives. Notice that the primitives are neither wrong nor right in themselves. They are certainly correct in some circumstances, and diSessa points out that experts use many primitives as quick and easy bits of compiled knowledge—but they are linked so as to only be used in appropriate circumstances.

I like to add an additional structure. Some of diSessa's phenomenological primitives are very abstract (*Ohm's primitive*, for example), and others refer to reasonably specific physical situations (*force as mover*, for example). I prefer to distinguish abstract reasoning primitives from those primitives applied in a particular context, which I refer to as *facets* (following Minstrell).[19] What I call an *abstract reasoning primitive* has a general logical structure, such as "if two quantities *x* and *y* are positively related, more *x* implies more *y*." What I call a facet implies a *mapping* of the slots in the primitive into particular variables in a particular physical context. This is illustrated in Figure 2.5. As diSessa points out [diSessa 1993], there are a very large number of facets, corresponding to the complexity of living in the world. In my formulation, this complexity is seen as arising from mapping a reasonably small number of abstract reasoning primitives (perhaps a few dozen) onto the large diversity of physical situations.

An example of mapping might be, "if the liquid is higher in one of the glasses, there is more liquid in that glass." In one of Piaget's classic experiments, children are shown a vessel containing some water. The water is then poured into a narrower vessel so that it rises to a higher level. Before the age of about five years, most children say that the amount of water has increased (because it's higher) or decreased (because it's narrower).[20] Both those children

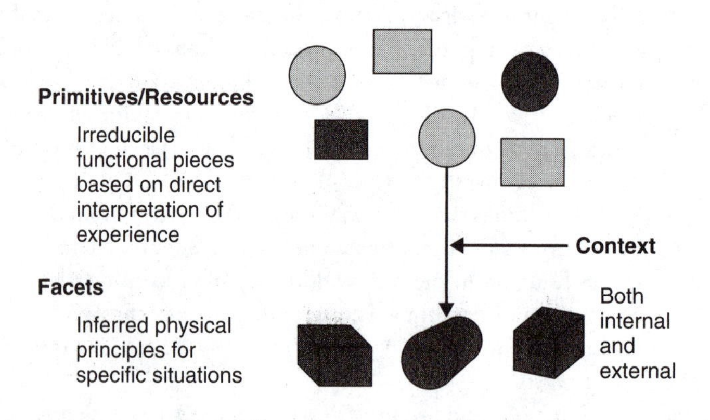

Figure 2.5 A visual representation of the mapping of abstract primitives into specific facets in a particular physical context.

[19] The term *facet* was introduced by Jim Minstrell [Minstrell 1992]. Minstrell listed large numbers of reasoning statements, some correct, others not, that students gave as explanations for their observations or in predictions about specific physical contexts.

[20] This is not a simple failure to understand the meaning of the question. A child may get upset that a sibling is "getting more" even if she is shown that when poured into a glass similar to their own they have the same heights.

who see it as more and those who see it as less are using what is essentially the same abstract reasoning primitive, but with different mappings (focusing on height or width). After the age of about five, children learn a "compensation" abstract reasoning primitive, something like "if two effects act on a variable, one to increase it and the other to decrease it, those effects tend to cancel." Children reaching this ability to reason are said to have achieved *Piagetian conservation.*

I've seen something very similar to this with engineers in first-year calculus-based physics at the University of Maryland. We were discussing the collision of two objects of different masses, and I asked about the relative size of the forces each object experiences. One group of students said, "The larger objects feel a bigger force since they have bigger masses." A second group said, "The smaller objects feel a bigger force since they change their velocity more." Only a small number of students had reached the "Piagetian conservation" stage of activating the compensation implicit in Newton's second law.

This kind of approach—analyzing the responses of our students in terms of primitives and facets—helps us understand more clearly the kinds of reasoning we can expect. The critical realization that arises from this kind of analysis is that students' common naïve conceptions are not simply wrong. They are based on correct observations but may be generalized inappropriately or mapped onto incorrect variables. If we can extract elements that are correct from students' common reasoning, we can build on these elements to help students reorganize their existing knowledge into a more complete and correct structure.

3. Activating resources from everyday experience: Situated cognition

The primitives discussed above tend to refer to specific real-world situations as asked about or observed in a physics class. A group of education specialists have focused on the difference between day-to-day reasoning and the kind of reasoning taught and learned in schools.

Sagredo once stopped me in the hall after his introductory physics class for physics majors. "You'll never guess what they couldn't do today! I was talking about projectile motion and asked them to describe what happens when a kicker kicks a football. I just wanted a description of the process—the ball goes up, travels a ways down the field, and comes down. No one would say *anything*, even when I pressed them. Why couldn't they give me a simple description?" I suspect, Sagredo, that it was because they weren't really sure what you wanted. They might well have expected that you wanted some technical or mathematical description in terms of forces, graphs, velocities, and accelerations. If they just said what you really wanted—the simple day-to-day physical description of the process—they were afraid they would look foolish. They may have been right to respond that way, given their previous experience with physics classes.

Most instruction in the United States today, despite reform efforts, continues to bear little relation to students' everyday lives. But many of the skills we are trying to teach can be tied to reasoning skills that the students possess and use every day. An interesting example comes from middle school arithmetic [Ma 1999]. Consider the following problem.

> A group of students has 3½ small pizzas, each whole divided into 4 parts. How many students can have a piece?

The reasoning used by a student to solve this problem is quite a bit different from the algorithm one learns for dividing fractions ($3\frac{1}{2}/\frac{1}{4} = 4 \times 3\frac{1}{2}$). The student might say something like: "Each pie can serve 4 students, so the 3 pies can serve 12. The ½ pie can serve 2, so a total of 14 can have a piece." This reasoning, like the reasoning we use to solve the Coke/Gin & Tonic problem in the previous section, relies on nonformal thinking that is linked to our everyday social experience. Tying it to the dividing by fractions problem—showing that division means finding how many times a part can be found in the whole—can help students make sense of what division really means and why division is a useful concept.

The use of context knowledge to help solve problems is a common feature of how people reason their way through situations in everyday life. A group of educators led by Jean Lave and Lucy Suchman [Lave 1991] [Suchman 1987] places this cognitive fact at the center of their educational reform efforts, creating *cognitive apprenticeships* and using *situated cognition.* There is an extensive educational literature on this topic,[21] and some dramatic improvements have been gained in children's understanding of and effectiveness using arithmetic by finding ways to use these everyday resources.

IMPLICATIONS OF THE COGNITIVE MODEL FOR INSTRUCTION: FIVE FOOTHOLD PRINCIPLES

Any model of thinking is necessarily complex. We think about many things in many ways. In order to find ways to see the relevance of these cognitive ideas and to apply them in the context of physics teaching, I have selected five general principles that help us understand what happens in the physics classroom.

1. The constructivism principle
2. The context principle
3. The change principle
4. The individuality principle
5. The social learning principle

1. The constructivism principle

> Principle 1: Individuals build their knowledge by making connections to existing knowledge; they use this knowledge by productively creating a response to the information they receive.

This principle summarizes the core of the fundamental ideas about the structure of long-term memory and recall. The basic mechanism of the cognitive response is context-dependent association. A number of interesting corollaries, elaborations, and implications that are relevant for understanding physics teaching come from the constructivism principle.[22]

[21] See, for example, [Lave 1991]. A very readable introduction to the subject is [Brown 1989].

[22] These properties point out that our whole structure of patterns of association/schemas/mental models is a somewhat fuzzy one. The boundaries between the different structures are not sharply delineated. There are lots of examples of this sort of description in physics. Consider, for example, the excitations of a crystal lattice. We can describe the excitations in terms of phonons or in terms of continuous waves. In some limiting cases, it is clear which description is the more useful; in others, they may overlap.

Some of the characteristics of schemas clarify what is happening when students make mistakes. Often in listening to my students explain what they think, I used to become confused and sometimes irritated. How can they say *x* when they know the contradictory principle *y*? Why can't they get started on a problem when they certainly know the relevant principle? They just told it to me two minutes ago! Why did they bring up that particular principle now? It doesn't apply here! The well-documented characteristics of mental structures described in the first part of this chapter help us understand that these sorts of errors are natural and to be expected.

It also makes us realize that we must get a lot more feedback than we traditionally get if we want to understand what our students are really learning. Traditional testing often fails to show us what students really think or know because many different schemas can produce the correct solution to a problem. Even if a student goes through the same steps as we do, there's no guarantee that their schema for choosing the steps is the same as ours.[23] I once asked a student, who had done a homework problem correctly, to explain his solution. He replied: "Well, we've used all of the other formulas at the end of the chapter except this one, and the unknown starts with the same letter as is in that formula, so that must be the one to use."

Part of the way we fool ourselves with standard testing methods is that we are interested "in what students know." If they don't access the right information in an exam, we give them clues and hints in the wording to activate access. But since essential components of a schema are the processes for access to information, we are not testing the students' patterns of associations if we narrow the context and provide detailed cues. The student "has" the information, but it is inert and cannot be used or recalled except in very narrow, almost preprogrammed situations.

To find out what our students really know, we have to find out what resources they are bringing and what resources they are using. We have to give them the opportunity to explain what they are thinking in words. We must also only give exam credit for reasoning and not give partial credit when a student tries to hit the target with a blast of shotgun pellets and accidentally has a correct and relevant equation among a mass of irrelevancies. To know whether our students access the information in appropriate circumstances, we have to give them more realistic problems—problems that relate directly to their real-world experience and do not provide too many "physics clues" that specify an access path for them. (I'll discuss the implications of this for assessment in chapter 4.)

2. The context principle

The second principle reminds us of the nonuniqueness of the cognitive response and sets the stage for the description of the dynamics of building mental structures.

> _Principle 2:_ What people construct depends on the context—including their mental states.

It's very easy to forget the warnings and drop back into the model that assumes students either know something or don't. Focusing on the context dependence of a response helps us

[23] The difficulty is that the mapping from underlying schema to problem-solving steps is not one-to-one. A specific example of this is given in [Bowden 1992].

keep in mind that the situation is not that simple. Nice examples of context dependence in students' responses to physics abound (although they are sometimes not presented that way). One particularly clear example comes from the work of Steinberg and Sabella [Steinberg 1997] [Sabella 1999]. At the end of the first semester of engineering (calculus-based) physics at the University of Maryland, they gave a section of 28 students the (to an expert) equivalent pair of questions, shown in Figure 2.6.

The first question is taken from the Force Concept Inventory (FCI) [Hestenes 1992].[24] It is stated in common speech using familiar everyday objects. The second is a typical physics class situation. It is abstract, and it involves idealized laboratory-style objects. The FCI was given as an ungraded diagnostic at the end of classes[25] and the problem as part of the final exam one week later. Both test the students' understanding of Newton's first law.

Although 25 (~90%) got the exam question correct, only 15 of the students (~55%) got the FCI question right. Nearly half of the students who succeeded with the problem in the exam context missed the question presented in a nonphysics context (11/25 ~ 45%). Interviews with the students suggest that this is a real phenomenon, not just a result of an additional week's time to study.

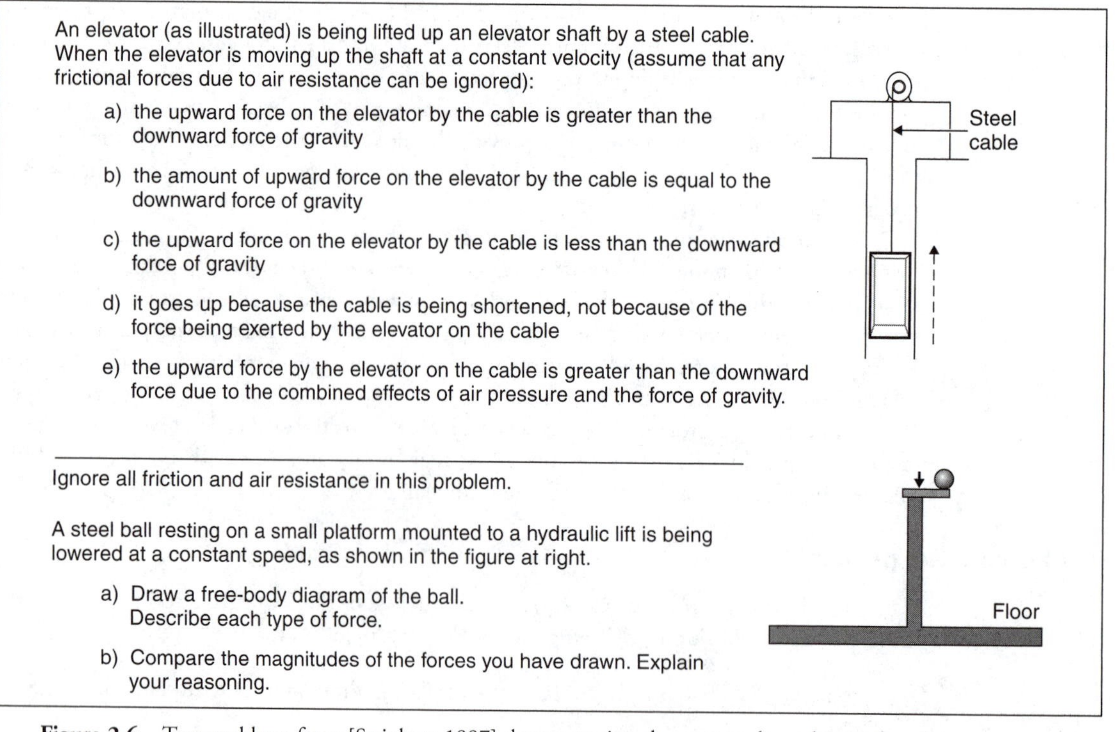

Figure 2.6 Two problems from [Steinberg 1997] demonstrating the context dependence of student responses.

[24] The FCI is discussed in detail in chapter 5 and is included in the Resource CD associated with this volume.

[25] These tests are often given as ungraded in order to encourage the students to give the answer they believe rather than the answer they might think we want in physics class.

What we really want to help our students do is build their knowledge into a coherent schema that is appropriately linked and that is triggered in a wide range of appropriate contexts.

3. The change principle

This principle deals with the dynamics of the mental state. It states that schemas are not only the way that we organize our interactions with the world, but they also control how we incorporate new information and experiences [Bransford 1973].

> Principle 3: It is reasonably easy to learn something that matches or extends an existing schema, but changing a well-established schema substantially is difficult.

I pose a number of restatements and elaborations of this principle as corollaries to clarify what it means for teaching.

> Corollary 3.1: It's hard to learn something we don't almost already know.

All students have things they know (some of which may be wrong!), things they are a bit familiar with, and things they have no knowledge about at all.

I like to look at this as an archery target. What they know is the bull's-eye—a compact black area; what they know a little about is a gray area surrounding the black; outside that is a white "rest of the world" about which they are clueless. To teach them something, we do best to hit in the gray. A class full of students is a challenge because all of their gray areas are not the same. I want to hit as many of the grays as possible with each paint-tipped shaft of information to turn gray areas black.[26] This metaphor only describes some aspects of the process. The real issue is that when we "hit in the gray," the student has many appropriate links that can be made to weave the new knowledge into their existing structure in an appropriate way.

In communication studies, an important implication of this corollary is called the *given-new principle* [Clark 1975] [RedishJ 1993]. It states that new information should always be presented in a context that is familiar to the reader and that the context should be established first. The analogous statement is very important in physics teaching, especially at the introductory level. As physicists with years of training and experience, we have a great deal of "context" that our students don't possess. Often we are as fish in water; unaware of having this context and unaware that it is missing in our students.

We can cite a number of specifics that are violations of the given-new principle. One important example is that we often use terms that students are not familiar with—or that they use in a different sense than we do. Lakoff and Johnson [Lakoff 1980], as a part of their study of the way speakers of English build their meaning of the term *force,* classified the characteristics of common metaphors using the term. Among their list of 11 characteristics, 8 involved

[26] This picture also interacts strongly with the social learning principle discussed below. Items in the gray are those that the student can learn via social interactions with teachers or more expert students. The followers of the Russian psychologist Lev Vygotsky give the gray region the unfortunate name "zone of proximal development" [Vygotsky 1978].

the will or intent of an individual! But most of us are so familiar with the technical meaning of "force" that we are surprised to learn that a significant fraction of our introductory students do not believe that a table exerts a force on a book it is supporting [Minstrell 1982]. Why doesn't the book fall through? The table is just "in the way." (This issue is discussed in more detail under the heading "Bridging" later in this chapter.)

The problem caused by the interpretation of common speech words for technical ones is not simple. I know that the terms "heat" and "temperature" are not really distinguished in common speech and are used interchangeably for the technical terms "temperature" (average energy per degree of freedom), "internal energy", and "heat flow" (flow of internal energy from one object to another). In one class, I stated this problem up front and warned my students that I would use the terms technically in the lecture. Part way through I stopped, realizing that I had used the word "heat" twice in a sentence—once in the technical sense, once in the common speech sense.[27] It's like using the same symbol to stand for two different meanings in a single equation. You can occasionally get away with it,[28] but it really isn't a good idea!

Putting new material in context is only part of the story. Our students also have to see the new material as having a plausible structure in terms of structures they know. We can state this as another useful corollary.

Corollary 3.2: Much of our learning is done by analogy.

This and the previous corollary make what students know at each stage critical for what we can teach them. Students, like everyone else, always construct their knowledge, and what they construct depends on how what we give them interacts with what they already have. This has important implications for the development of instructional techniques that help students overcome strong misconceptions. (See the discussion of "bridging" in a later section.)

One implication of these results is that we should focus on building structures that are useful for our students' future learning. I state this as a third corollary.

Corollary 3.3: "Touchstone" problems and examples are very important.

By a *touchstone problem*,[29] I mean one that the student will come back to over and over again in later training. Touchstone problems often become the analogs on which they will build the more sophisticated elements of their schemas. It becomes extremely important for students to develop a collection of a few critical things that they really understand well.[30] These become the "queen bees" for new swarms of understanding to be built around. I

[27] "If there is no heat flow permitted to the object, we can still heat it up by doing work on it."

[28] I have seen colleagues write the energy levels of hydrogen in a magnetic field as $E_{nlm} = E_n - \left(\frac{e\hbar}{2m}\right) mB$ where the m in the denominator is the electron mass and the one in the numerator is the z-component of the angular momentum. Most physicists can correctly interpret this abomination without difficulty.

[29] In his discussion of scientific paradigms, T. S. Kuhn refers to these problems as *exemplars* [KuhnT 1970].

[30] In addition to giving them centers on which to build future learning, knowing a few things well gives the student a model of what it means to understand something in physics. This valuable point that has been frequently stressed by Arnold Arons [Arons 1990]. It is an essential element in developing scientific schemas.

believe the sense that some superficially uninteresting problems serve this role is the reason they have immense persistence in the community. Inclined plane problems really aren't very interesting, yet the occasional suggestions that they be done away with are always resisted vigorously. I think the resisters are expressing the (often unarticulated) feeling that these are the critical touchstone problems for building students' understanding of vector analysis in the plane.

Corollary 3.3 is one reason we spend so much time studying the mass on a spring. Springs are of some limited interest in themselves, but small-amplitude vibrations are of great general importance. The spring serves as a touchstone problem for all kinds of harmonic oscillations from electrical circuits up to quantum field theory.

Analyzing a curriculum from the point of view of the schema we want students to develop, their preexisting schemas, and touchstone problems that will help them in the future can help us understand what is critical in the curriculum, which proposed modifications could be severely detrimental, and which might be of great benefit.

Combining these ideas with the idea of associations discussed under Principle 1 leads us to focus on the presence of a framework or structure within a course. It suggests that building a course around a linked series of touchstone problems could be of considerable assistance in helping students understand the importance and relevance of each element. Such a structure is sometimes referred to as a *story line.*

Unfortunately, if students are *not* blank slates, sometimes what is written is—if not entirely wrong—inappropriate for future learning in physics.[31] Then it can seem as if we have run into a brick wall. This brings us to the next corollary.

Corollary: 3.4: It is very difficult to change an established mental model.

Traditionally, we've relied on an oversimplified view of Principle 1, the constructivism principle, to say: "Just let students do enough problems and they'll get the idea eventually." Unfortunately, this simple translation of the principle doesn't necessarily work. Although practice is certainly necessary to help students compile skills into easily retrievable knowledge, there is no guarantee that they will link those skills into a structure that helps them to understand what's going on and how to use the basic concepts appropriately.

The limitations of doing lots of problems were investigated in a study done in Korea. Eunsook Kim and Jae Park looked at the response of 27 students in an introductory college physics class to the FCI [Kim 2002]. American students at large, moderately selective state universities (such as the University of Maryland, the Ohio State University, or the University of Minnesota) who have taken one year of high school physics score an average of 45–50% on this test before beginning a calculus-based physics class. The students Kim studied had taken an apparently much more rigorous high school physics program in which each student had done an average of about 1500 problems (ranging between 300 and 2900) end-of-chapter problems. In a typical American high school class, students will do 300 to 400 such problems. Despite doing 5 to 10 times as many problems as American students, the students

[31] Perhaps a *palimpsest* is a better metaphor for a student's mind than a blank slate. According to the American Heritage Dictionary, a palimpsest is "a manuscript, typically of papyrus or parchment, that has been written on more than once, with the earlier writing incompletely erased and often legible."

still had substantial conceptual difficulties with fundamental concepts of mechanics at rates comparable to those seen with American students. There was little correlation between the number of problems the students had done and the conceptual understanding displayed.

This study and others like it are a warning against relying on the idea that "repetition implies effective learning"—that is, that frequent repetition of a particular type of activity is *all* that is needed to produce a useful functional schema. Repetition *is* necessary to create compiled knowledge, but it is not sufficient. For effective usage, the compiled element needs to be linked into a coherent schema about the subject.

Once students learn how to do problems of a particular type, many will learn nothing more from doing more of them: New problems are done automatically without thinking. This also means that testing by varying homework problems slightly may be inadequate to probe the student's schemas. More challenging tests involving a variety of modes of thinking (problem solving, writing, interpreting, organizing) are required. Such testing is discussed in detail in chapter 4.

It has been demonstrated over and over again that simply telling students some physics principle doesn't easily change their deep ideas.[32] Rather, what often happens is that instead of changing their schema substantially, a poorly linked element is added with a rule for using it only in physics problems or for tests in one particular class. This and the fact that a schema can contain contradictory elements is one possible reason "giving more problems" is often ineffective.

A few years ago, I heard a lovely anecdote illustrating the barriers one encounters in trying to change a well-established mental model.[33] A college physics teacher asked a class of beginning students whether heavy objects fall faster than light ones or whether they fall at the same rate. One student waved her hand saying, "I know, I know." When called on to explain she said: "Heavy objects fall faster than light ones. We know this because Galileo dropped a penny and a feather from the top of the leaning tower of Pisa and the penny hit first." This is a touchstone example for me. It shows clearly that the student had been told—and had listened to—both the Galileo and the Newton stories. But she had transformed them both to agree with her existing mental model.[34]

Fortunately, mechanisms are available for helping students restructure even well-established mental models; these methods are discussed later in this chapter.

4. The individuality principle

One might be tempted to say: Fine. Let's figure out what the students know and provide them with a learning environment—lectures, demonstrations, labs, and problems—that takes them from where they are to where we want them to be. Since we all know that a few

[32] See the papers referred to in the annotated bibliography [McDermott 1999] and in the review papers [McDermott 1991], [Reif 1994], and [Redish 1999].

[33] Audrey Champagne, private communication.

[34] We should not lose sight of the fact that the student's mental model in this case is in fact correct. We observe that lighter objects *do* fall more slowly than heavy ones if they fall in air, and few of us have much direct experience with objects falling in a vacuum. But for reasonably dense objects falling for only a few seconds, the difference is small, so that this observation does not yield a useful idealization. The observation that objects of very different mass fall in very nearly the same way does.

students get there from here using our current procedures, why can't we make it work for all of them? We do in fact know now that the right environment can produce substantially better physics learning in most of the students taking introductory university physics.[35] But my fourth principle is a word of warning that suggests we should not be looking for a "magic bullet."

> *Principle 4:* Since each individual constructs his or her own mental structures, different students have different mental responses and different approaches to learning. Any population of students will show a significant variation in a large number of cognitive variables.

I like to call this principle the *individuality* or *distribution function principle.* This reminds us that many variables in human behavior have a large natural line width. The large standard deviation obtained in many educational experiments is not experimental error; it is part of the measured result! As physicists, we should be accustomed to such data. We just aren't used to its being so broad and having so many variables. An "average" approach will miss everyone because no student is average in all ways.[36]

In addition to the fact that students have different experiences and have drawn different conclusions from them, their methods of approach may differ significantly. I state this as a corollary.

> *Corollary 4.1:* People have different styles of learning.

There is by now a vast literature on how people approach learning differently. Many variables have been identified on which distributions have been measured. These include authoritarian/independent, abstract/concrete, and algebraic/geometric, to name a few. The first variable means that some students want to be told, while others want to figure things out for themselves. The second means that some students like to work from the general to the specific, and some the other way round. The third means that some students prefer to manipulate algebraic expressions, while others prefer to see pictures. Many of us who have introduced the computer in physics teaching have noted that some students want to be guided step by step; others explore everything on their own. These are only a few of the variables. For some good analyses of individual cognitive styles and differences, see [Gardner 1999], [Kolb 1984], and [Entwistle 1981].

Once we begin to observe these differences in our students, we have to be exceedingly careful about how we use them. A preference does not mean a total lack of capability. Students who prefer examples with concrete numbers to abstract mathematical expressions may be responding to a lack of familiarity with algebra rather than a lack of innate ability. Many of our students' preferences come from years of being rewarded for some activities (such as

[35] As examples, see the references cited in the section "Research-Based Instructional Materials" in [McDermott 1999]. This annotated bibliography is also included on the Resource CD.

[36] This is analogous to the story of the three statisticians who went hunting deer with bows and arrows. They came across a large stag and the first statistician shot—and missed to the left. The second statistician shot and missed to the right. The third statistician jumped up and down shouting "We got him!"

being good memorizers) and chastised for others (such as asking questions the teacher couldn't answer). Expanding our students' horizons and teaching them how to think sometimes requires us to overcome years of negative training and what they themselves have come to believe are their own preferences and limitations.

An important implication is the following:

> Corollary 4.2: There is no unique answer to the question: What is the best way to teach a particular subject?

Different students will respond positively to different approaches. If we want to adopt the view that we want to teach all our students (or at least as many as possible), then we must use a mix of approaches and be prepared that some of them will not work for some students. We need to answer the question: What is the distribution function of learning characteristics that our students have in particular classes? Although some interesting studies have been done over the years, the implication for instruction in physics is not well understood.[37]

Another implication that is very difficult to keep in mind is:

> Corollary 4.3: Our own personal experiences may be a very poor guide for telling us the best way to teach our students.

Physics teachers are an atypical group. We "opted in" at an early stage in our careers because we liked physics for one reason or another. We then trained for up to a dozen years before we started teaching our own classes. This training stretches us even farther from the style of approach of the "typical" student. Is it any wonder that we don't understand most of our beginning students and they don't understand us?

I vividly recall a day a few years ago when a student in my algebra-based introductory physics class came in to ask about some motion problems. I said: "All right, let's get down to absolute basics. Let's draw a graph." The student's face fell, and I realized suddenly that a graph was not going to help him at all. I also realized that it was going to be hard for *me* to think without a graph and to understand what was going through the student's mind. I never minded doing without derivatives—motion after all is the study of experimental calculus, and you have to explain the concept (maybe without using the word "derivative") even in a non-calculus-based class. But I can't remember a time when I couldn't read a graph, and I have found it difficult to empathize with students who come to physics and can't read a graph or reason proportionately. It takes a special effort for me to figure out the right approach.

This is very natural given the earlier principles. Our own schemas for how to learn come from our personal reactions to our own experiences. However, to reach more of our students than the ones who resemble ourselves, we will have to do our best to get beyond this mindset. It makes the following principle essential.

> Corollary 4.4: The information about the state of our students' knowledge is contained within them. If we want to know what they know, we not only have to ask, we have to listen!

[37] See [Kolb 1984].

One point I want to stress about the individuality principle is in the idea expressed by its final sentence: *Any population of students will have a significant variation in a large number of cognitive variables.* We have a tendency, especially in physics, to classify students along a single axis that we think of as "intelligence." I've heard Sagredo say, "Well, most of my students are having trouble, but the smart ones get it." In cognitive science, there have been vigorous arguments for a number of years now as to whether there is a single variable (referred to as "g") that describes intelligence, or whether what we call intelligence consists of a number of independent factors.[38] The literature on this subject is quite complex, and I do not pretend to be able to evaluate it. However, whether or not intelligence is a unary concept, success in physics—or in any scientific career—relies on much more than intelligence. I have followed with interest the careers of many of my physics classmates from college and graduate school for many decades, and one thing is absolutely clear. The students who were the "brightest" in doing their schoolwork were not necessarily the ones who went on to make the most important contributions to physics. Creativity, persistence, interpersonal skills, and many other factors also played large roles.[39] This point is discussed again later in the next chapter on the hidden curriculum.

5. The social learning principle

With the fifth principle, I go beyond single individuals and consider their relations to others as a part of their learning. This principle is based on the work on group learning that builds on the ideas of the Russian psychologist, Lev Vygotsky. These ideas have had a profound impact on modern theories of teaching and learning [Vygotsky 1978] [Johnson 1993].

> Principle 5: For most individuals, learning is most effectively carried out via social interactions.

The social learning principle is particularly important for physicists to keep in mind. Physicists as a group are highly unusual in many ways. In my experience, they tend to be in the extreme tails of distributions of curiosity, independence, and mathematical skills. They also tend to be highly self-sufficient learners. I once heard David Halliday remark that what he enjoyed most as a student was sitting down by himself alone in a quiet room with a physics text and going "one-on-one" with the authors of the book—trying to understand them and figure out what they were saying. Many of us have similar inclinations. Physicists as a group seem to be selected for the character of being able to learn on their own. But in examining my experiences of this type, I have decided that my "learning on my own" involves an ability to create an "internalized other"—to take a variety of viewpoints and to argue with myself. This is not a commonly found characteristic and should not be assumed in a general population of students.

[38] For a discussion and for references to this literature, see [Gardner 1999].

[39] The range of important variables provides a basis for what we might call *Garrison Keillor's Corollary.* "All students are above average—on some measure."

SOME GENERAL INSTRUCTIONAL METHODS DERIVED FROM THE COGNITIVE MODEL

Many instructional methods have been developed based on the cognitive models discussed above. Two that will be relevant for us later in our discussions of specific curricula are *cognitive conflict* and *bridging*. The cognitive conflict method is used when an inappropriate generalization or incorrect association has become particularly robust and difficult to change. The bridging method relies on the explicit idea that the students bring useful knowledge resources to their learning of physics and attempts to explicitly activate those resources in appropriate ways.

Cognitive Conflict

Common naïve conceptions can be strikingly robust. At the beginning of my unit on direct current circuits in my calculus-based engineering physics class, I gave the students the problem shown in Figure 2.7.

This problem is by now rather traditional. It appears in sixth grade science texts and in many high school physics texts. My discussion of circuits occurred at the beginning of our third semester of engineering physics. Most of the students were second-semester sophomores, and many of them were electrical engineering students. More than 95% of them had taken (and done well in) high school physics. Yet only about 10-15% of them were able to solve this problem before instruction.

Sagredo complained that this was "a trick question." He said, "I'll bet many graduate students in physics would miss it. You have to find a clever way to make it light—touching the bulb's tip to one end of the battery to make a contact without a wire. Then you can use the wire to close the circuit."

Sagredo's correct answer is shown at the left of Figure 2.8. He is right that many physics grad students will miss the problem on the first try (and many professors as well), but the specific wrong answers given by students and by experts show that something different is going on. The experts who get it wrong give the answer, "No. It can't be done. You need a closed circuit and that takes two wires." My engineering students' wrong answers were, "Sure you

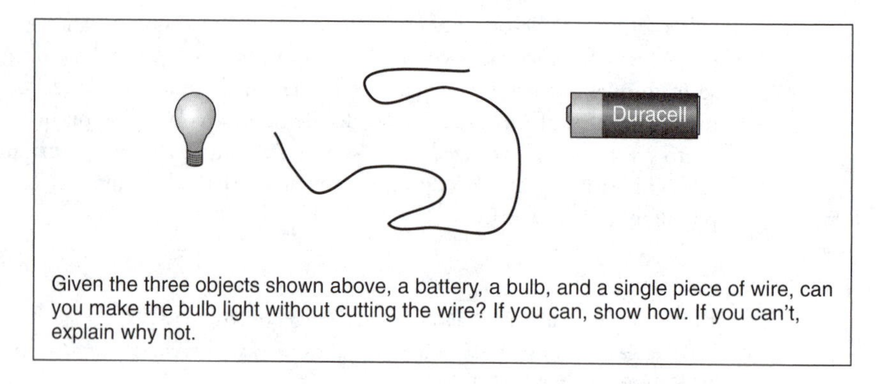

Figure 2.7 A problem introductory students often have difficulty with.

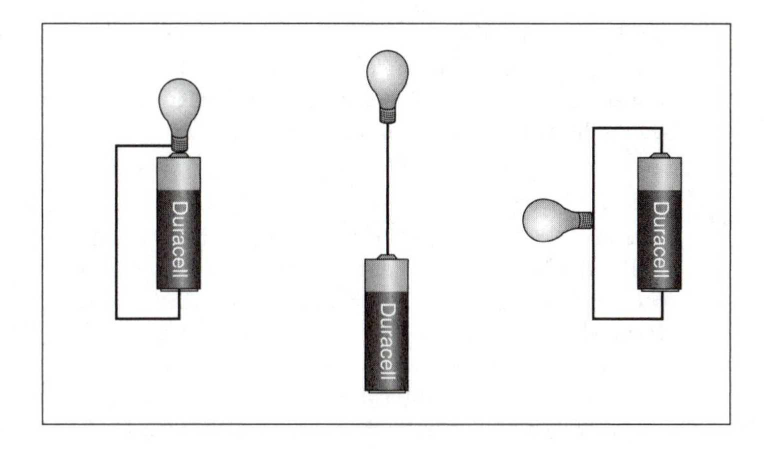

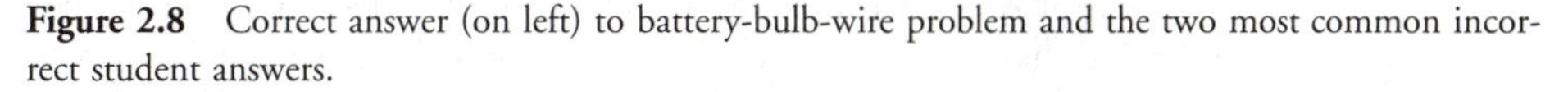

Figure 2.8 Correct answer (on left) to battery-bulb-wire problem and the two most common incorrect student answers.

can. It's easy." One-third of the students gave the answer shown in the middle, and one-third gave the answer shown at the right. About half of the rest gave the correct answer. The rest either left it blank or gave uninterpretable answers. Very few students gave the "expert's error."

Many students showed the common naïve conception of electricity as a source of energy that can be "drawn" directly from an electric power source or "tapped into" by touching. Students' naïve conceptions about electric currents are documented in the research of the University of Washington Physics Education Group (UWPEG) [McDermott 1992]. The group also reports a lesson they have developed to address this issue [Shaffer 1992]. The lesson is delivered in a "Tutorial"—a structure that has replaced recitations at the University of Washington. These and related materials are a part of the Physics Suite and are discussed in detail in chapter 8.

The model frequently used in Tutorials is cognitive conflict in the form *elicit/confront/resolve/reflect.* In the first Tutorial on direct currents, the lesson begins with the question shown above (given on a pretutorial ungraded quiz[40] during lecture). When the students get to the Tutorial, each group gets a battery, a bulb, and a single wire.

When I gave this lesson in my classes, about half of the students expected to be able to light the bulb using one of the two arrangements at the right of Figure 2.8. I particularly remember one student who came up to me complaining that the equipment she had been given had to be defective. (She was certain that the middle arrangement shown in Figure 2.8 should work. "After all," she said, "that's the way a flashlight works, isn't it?") She insisted on having a new bulb and a fresh battery.

The subsequent discussions with their colleagues and with the facilitators (and the rest of the lesson, which elaborates on the point and reconsiders it in a variety of contexts) help students to resolve the conflict between their model and their observations, and to reflect

[40] One reason for not grading the quiz is to encourage students to look for what they actually think is plausible rather than to try to guess what the teacher wants. The solutions also are not posted since the point is to get students thinking about the reasoning (during the Tutorial period) rather than focusing on the answer.

upon the implications. After Tutorial instruction, students did significantly better (~75% success) on other (more complex) questions probing similar issues than students with traditional instruction (< 50% success). Many other examples exist of successful learning produced through lessons based on cognitive conflict.

Bridging

The use of cognitive conflict as the primary instructional tool in the classroom can lead to difficulties because it is a rather negative approach. A colleague who had instituted a reform procedure in a high school physics class that relied heavily on the cognitive-conflict method reported to me that one of his students responded to a quiz: "Oh great. Here's another test to show us how stupid we are about physics." Once students are convinced they cannot do physics, it is extremely difficult to teach them anything more than rote memorization.[41]

However, if we consider the student's starting knowledge as a resource and begin with what is *right* about student reasoning, we can make them feel considerably better about themselves and about their ability to do physics. To see how this might work, let's consider an example.

John Clement has proposed building on the correct aspects of students' naïve conceptions by finding a series of *bridges* or *interpolating steps* that help students transform their mental models to match the accepted scientific one [Clement 1989]. The starting point for a bridge should be something that the students know well that is correct—which Clement refers to as an *anchor*. An example of bridging is in Clement's building of a lesson to respond to the common naïve conception about normal forces.

In his classic paper, "Explaining the 'At Rest' Condition of an Object," Jim Minstrell documented that more than half of the students in his high school physics class did not believe that an inanimate supporting object could exert a force [Minstrell 1982]. They interpreted what a table does to a book resting upon it as "simply preventing it from falling." They showed that the pattern of association triggered by the word "force" had a strong link to the idea of will or intent and hence, to something produced by an "active" object, most frequently an animate one. Students seem to be bringing up a "blocking" reasoning primitive rather than a "springiness" one. This result has subsequently been confirmed to occur in a significant fraction of introductory physics students.

Clement looked for examples in which students had correct intuitions (primitive responses or "gut feelings") that could be used to build on. He suggests that a useful starting point for a bridging lesson should have the following characteristics.

1. It should trigger with high probability a response that is in rough agreement with the results of correct physical theory.
2. It should be concrete rather than abstract.
3. The student should be internally confident of the result; that is, students should strongly believe in the reasoning by themselves and not by virtue of authority.

Clement came up with two possible anchors for the situation described by Minstrell.

- Holding up a heavy book in your hand
- An object being held up by a reasonably soft spring

[41] These issues are discussed in more detail in chapter 3.

TABLE 2.1 Results of a Bridging Lesson

	Pre-test average	Post-test average	Fraction of possible gain achieved
Control group ($N = 55$)	(17 ± 1)%	(45 ± 2)%	0.34
Experimental group ($N = 155$)	(25 ± 2)%	(79 ± 1)%	0.72

To Clement's surprise, the spring served as a much better starting point than the book. This stresses the fact that we cannot assume that an example that seems an obvious anchor to us will work for our particular population. They must be tested. Our individuality principle also reminds us not to expect that a particular example will serve for all students. We may have to use a number of different anchors to reach most students.

Clement tested this approach in four high schools in Massachusetts [Clement 1998]. There were 155 students in the experimental group and 55 in the control. The same pre- and post-tests were given to all students. On the issue of static normal forces, the gains of the groups were as shown in Table 2.1. The experimental group was taught with a bridging lesson; the control class with a traditional one. Evaluation was done with six questions that were part of a longer test. (Errors are std. error of the mean.) Similar results were reported for bridging strategies in the areas of friction and dynamic use of Newton's third law in collisions.[42]

Restricting the frame

The real world, with all its complexity, is too much for us to handle all at once. Long-term and working memories function together to parse the visible world into the pieces relevant for functioning in any particular situation. Many students, when facing physics problems, have considerable difficulty figuring out what is important and what is not. Helping them build appropriate templates and associations is an important part of what physics instruction is trying to accomplish.

Limiting our view of what we want to analyze is an essential part of the scientific enterprise. One of Galileo's greatest contributions to science was his ability to step back from the Aristotelian attempt to produce a grand synthesis for all motion and to focus on understanding well how a few simple phenomena really worked—an object on an inclined plane, the pendulum, a falling body. The synthesis comes later, once you have some solid bricks to build into a more coherent structure. When Newton synthesized a theory of motion 50 years later, he created a limited synthesis—a theory of motion, but not a single theory that successfully described, for example, light, heat, and the properties of matter at the same time. The scientific structure grows at different paces and in different places, with pieces being continually matched and modified to create more coherent and useful maps.

Similarly, in teaching introductory physics, we also have to restrict our considerations to a piece of the entire picture. Our goal is not to present the most coherent picture of the entire physical knowledge system that we as professionals have been able to construct, even

[42] For a discussion of the lessons and the controls, see [Clement 1998].

though this might be an intellectually enticing and enjoyable goal. The mental structure that we've created for ourselves to describe physics has been built up over many years—perhaps decades—and has resulted from continual transformations that have been made not only to the physics, but to ourselves.

Even in solving a single problem in introductory physics, we must go through the process of restricting consideration to a piece of the world and of limiting those aspects we want to consider. I refer to selecting a limited piece of the world to view through a frame as *cat television.* My cat (see Figure 2.9) very much enjoys viewing a small piece of the world through a window (as have many other cats I have known) and can become quite addicted to it. When we drove across the country for a sabbatical visit with the cat in the car, in every motel room he insisted on finding a place where he could look out the window and see where in the world we were. The real TV doesn't interest him at all. (The same is true for many physicists.)

Once we have drawn our frame—chosen our channel on cat TV—we still have work to do before we can start drawing a representation. We have to choose what to pay attention to and what to ignore. I call this process *creating the cartoon.* I mean "cartoon" here in the sense of an artist's sketch, drawn in preparation to creating a painting or mural. The process of deciding what to keep and what to ignore is a difficult one, and one that is often glossed over by physicists who already know what to consider. (An example is Bill Amend's FoxTrot cartoon—in the other sense—shown in Figure 2.10.)

Changing what one decides to look at in a real-world physical phenomenon can be associated with a major advance in science. In his *Structure of Scientific Revolutions,* T. S. Kuhn [KuhnT 1970] describes the paradigm shift associated with Galileo's observation of the pendulum. The story goes that Galileo was sitting in church, bored with an extended sermon, perhaps, when a gust of wind started a chandelier swinging. An observation that could be made on the chandelier is that it eventually comes to rest. One could infer that this supports the Aristotelian position that objects naturally come to rest (and seek the lowest point). But Galileo timed the swing of the chandelier using his pulse and noted that the period of the

Figure 2.9 "Cat television." Cats often enjoy observing a small, framed "piece of the world." Similarly, a scientist must narrow his or her focus in a particular set of issues in order to make progress. I call this "choosing a channel on cat TV."

Figure 2.10 Assumptions in traditional physics problems are often tacit but important.

swing did not change as the amplitude got smaller. He then realized that this could be interpreted as saying that the natural state of the object was to continue to oscillate. In the Aristotelian view, the coming to rest is taken as fundamental, and the oscillation is seen as a deviation. In the Galilean view, the oscillation is seen as fundamental, and the damping is seen as peripheral. There is no *a priori* way to decide which view is better. One must have more information as to how these assumptions play out and how productive they are in generating deeper understandings of a wider class of phenomena.

When students view a subject or a problem in physics, they are often unclear as to how to link what the physicist sees in the problem with their personal experience. A real-world problem involves friction, massive pulleys, cartwheels that have significant moments of inertia, etc., etc., etc. *Knowing what to ignore is an important part of learning to think about science, and it should not be treated as trivial.* Sagredo often teaches his students "the first step in doing a physics problem is to draw a diagram," but expresses frustration that often the students "can't even do that." I have seen students who, when given that instruction, draw a 3-D diagram when a 2-D one will do, or who carefully draw a realistic car, person, or horse when the problem could be done by using a block. Is it obvious, when asked to "consider a car rolling down a hill," that it doesn't matter whether the car is a Porsche or a VW bug or a block of wood with wheels? I conjecture that our ignoring this step is one reason that students don't always see the relation of physics to their real-world experience (see chapter 3).

Multiple representations

In science we use a dizzying variety of representations[43] that may have a variety of positive cognitive effects. First, the use of multiple representations (visual and verbal, for example) may help us make better use of our working memory. Second, different representations associate more naturally with different features of the data/situation we are trying to describe. As a result, the use of multiple representations can be effective in building links between different aspects of the situation. Some of the representations we use in physics include

- words
- equations
- tables of numbers
- graphs
- specialized diagrams

For the expert, each representation expresses some characteristic of the real-world system more effectively than others, and the translation and links between them help build up a strong and coherent mental model. But each representation takes some time and effort to learn. If students haven't yet learned to "read" a representation, that is, if their knowledge of that representation hasn't been compiled into an easily processed chunk, the translation of the representation can take up too much working memory to allow the students to make effective use of the representation. Some may have mastered one or more of these representations through experiences they had before coming to our class. Some may have a learning style that favors a verbal or a visual representation. Some may think they have trouble with one or more of our representations and actively avoid thinking about them [Lemke 1990].

I represent our process of setting up a problem or scientific exploration with the diagram shown in Figure 2.11. We begin by picking a channel of cat television with a specific real-world situation considered in a specific physics context—for example, the mechanics of a car rolling down a hill. We then make a selection of what we want to look at first—a simplified image that corresponds to a box sliding down a frictionless inclined plane—what I called above: creating a cartoon. Is this a good model for an SUV rolling down a rocky mountain trail? Perhaps not. But it might be a good place to start for a car with smaller wheels rolling down a paved road.

Once we have our cartoon, we then express our knowledge of the situation in a variety of ways, using linked multiple representations. These present the information in different ways, enabling different ways of thinking about and linking the information.

There is considerable evidence in the research literature documenting student difficulties with representation translation [Thornton 1990] [Beichner 1994]. Students often see, for example, the drawing of a graph as the solution to a problem—something the teacher asked them to do—instead of a tool to help them understand and solve a more complex problem.

[43] See, for example, the detailed analysis of a chemistry and physics class in [Lemke 2000].

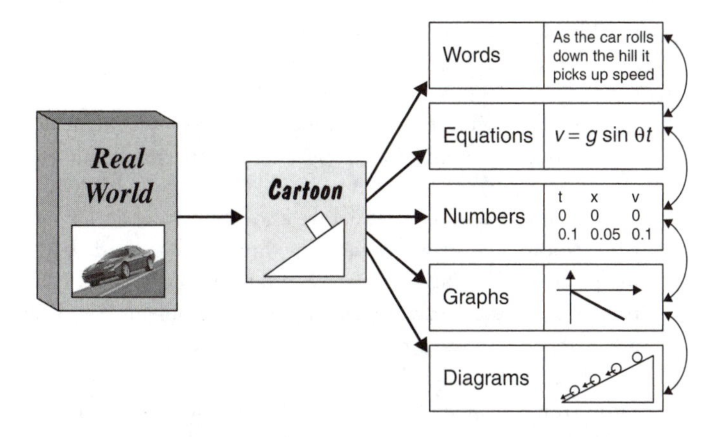

Figure 2.11 Physicists use a large number of different representations to describe events and processes in the real world. Learning to handle these representations and to translate them is an important—and difficult—step for introductory physics students.

In order to use multiple representations in our physics classes, it is important that we be aware of their difficulty for students and that we help those students both to learn the physics and to learn to make the connection to the modes of thinking they are less comfortable with.

RETHINKING THE GOALS OF PHYSICS INSTRUCTION

Putting our students' learning into a cognitive framework helps us begin a more detailed and constructive dialog in the community of physics teachers on what we want to accomplish with our physics instruction. There is more to learning physics than placing check marks on the table of contents of a text. Many of the most important results of our instruction are not associated with particular physics content. Even those goals that are associated with particular content can now be seen in a different way, given our understanding of student thinking. In this section, I try to explicate some of the "hidden" goals of attitudes and skills that we might like our students to attain in an introductory physics course—goals that are rarely discussed and whose attainment is often both strongly desired and taken for granted.

"Wait!" Sagredo interjects. "This talk of attitudes and skills is all very well, and I hope my students will develop them. But the physics content is important and we shouldn't lose sight of it." An excellent point, Sagredo. Let's start with what our learning theory says about learning physics content itself. Then we can consider attitude goals and skill development.

Extended content goals

Learning physics content means much more than memorizing a lot of independent definitions and equations. We want our students to understand enough about what the physics means to be able to understand what problems are about and what their answers imply; we want them to understand how the physics they are learning fits—both with other physics they have learned and with their personal experience with the world; and we want them to be able to use their knowledge effectively in solving problems. I refer to these three goals as *concepts*, *coherence*, and *functionality*.

> Goal 1: Concepts—Our students should understand what the physics they are learning is about in terms of a strong base in concepts firmly rooted in the physical world.

From our cognitive model we know that students can attach new knowledge to their existing knowledge structures as something separate and only relevant for the context of solving problems in a physics class. We want students to not only compile the mathematical manipulations associated with solving physics problems (i.e., learn to use them without thinking about them); we want them to understand what the physics is *about*.

To achieve this goal, students have to make sense of physics concepts—the ideas and definitions that map abstract physics descriptions onto real things in the physical world. In order to help them reach this goal through instruction, it often helps to motivate the need for a concept before introducing the definition through direct observation of phenomena. Arnold Arons called this "idea first, name afterwards" and stressed the value of operational definitions in physics instruction. He has a nice description of this in his book [Arons 1990].

> In my own courses, I indicate from the first day that we will operate under the precept "idea first and name afterwards," and that scientific terms acquire meaning only through the description of shared experience in words of prior definition. When students try to exhibit erudition (or take refuge from questioning) by name dropping technical terms that have not yet been defined, I and my staff go completely blank and uncomprehending. Students catch on to this game quite quickly. They cease name dropping and begin to recognize, on their own, when they do not understand the meaning of a term. Then they start drifting in to tell me of instances in which they got into trouble in a psychology or sociology, or economic, or political science course by asking for the operational meaning of technical terms. [Arons 1990]

While presenting "idea first, name second" is a good start, it's rarely sufficient to get students to develop a good understanding of physics concepts. A considerable amount of the effort in physics education research over the past decade has been devoted to the development of effective instructional techniques for helping students build their conceptual understanding. For more details on these specific content issues, see [Arons 1990] and the references in the Resource Letter [McDermott 1999] given on the resource CD.

> Goal 2: Coherence—Our students should link the knowledge they acquire in their physics class into coherent physical models.

A major strength of the scientific worldview is its ability to describe many complex phenomena with a few simple laws and principles. Students who emphasize science as a collection of facts may fail to see the integrity of the structure, an integrity that is both convincing and useful in solving problems. The lack of a coherent view can cause students many difficulties, including a failure to notice errors in their reasoning and an inability to evaluate a recalled item through cross-checks.

Let's recall Principles 1 and 2: Students will put what we give them into their knowledge structure and integrate it into their existing knowledge structure in some way of their own. Whatever it is that they know of any particular content, they may or may not create coherent schemas with appropriate connections that are activated with a high probability in appropriate contexts. We want our students to not simply "get the content" but to build their understanding of that content into an accurate and effective mental model.

> Goal 3: Functionality—Our students should learn both *how* to use the physics they are learning and *when* to use it.

For most of our populations of introductory physics students, we don't just want them to *know* the physics content but to be able to *do something with it.* My second cognitive principle suggests that in addition to having students master the physics content and see that it makes sense, we also want students' knowledge of physics to be robust and functional. That is, they should be able to recognize contexts that are appropriate for their particular physics knowledge and use them in correct ways. This means that we need to help students not only to obtain physics content knowledge, but to organize their knowledge.

These goals suggest that we should broaden our evaluation procedures. Traditionally we only test the content and part of the student's skill in doing physics, usually (at least at the introductory level) in limited preset contexts. Sagredo once decided that an effective way to get students to do their homework was to create exams made up of only previously assigned homework problems. The students were quite happy with this arrangement. Unfortunately, this sends the strong message that the only physics knowledge they need is restricted to a small set of problems whose solutions can be memorized. Students trained in this fashion are unlikely to be able to solve any other problems than the ones they have memorized. (I have seen situations where reversing the figure from left to right made the problem impossible for students who were quite comfortable solving the original problem.)

It is not sufficient for students to "know" the relevant correct statements of physics. They also have to be able to gain access to them at the appropriate times; and they have to have methods of cross-checking and evaluating to be certain that the result they have called up is truly relevant. To do this, they need to build a coherent and consistent mental model of the subject.

If we want to help our students build good models of the physics content, our cognitive model of student learning provides some guidance. The experience that outstanding teachers have reported is consistent with what we learn from researchers in cognitive and neuroscience: activities teaching thinking and reasoning have to be repeated in different contexts over a period of time. Arnold Arons observes:

> It must be emphasized, however, that *repetition* is an absolute essential feature of [effective] instruction—repetition *not* with the same exercises or in the same context but in continually altered and enriched context. . . . Experience . . . must be spread out over weeks and months and must be returned to in new contexts after the germination made possible by elapsed time. Starting with only a few students being successful, each repetition or re-cycling sees an additional percentage of the class achieving success, usually with a leveling of somewhat below 100% of the total after approximately five cycles. [Arons 1983]

Our cognitive colleagues tell us that to move something from short-term to long-term memory takes repetition and rehearsal. Our neuroscience colleagues are showing us that learning involves real structural changes among neural connections. I summarize this with a basic teaching and learning precept—really a corollary to the constructivism principle.

> *Corollary 1.1:* Learning is a growth, not a transfer. It takes repetition, reflection, and integration to build robust, functional knowledge.

This corollary leads to a guideline for instruction. I refer to it (and to subsequent guidelines for instruction) as a "teaching commandment." The full set (along with the cognitive principles and goals listed in this chapter and elsewhere) are summarized in a file on the Resource CD.

> *Redish's first teaching commandment:* Building functional scientific mental models does not occur spontaneously for most students. They have to carry out repeated and varied activities that help build coherence.

CHAPTER 3

There's More Than Content to a Physics Course: The Hidden Curriculum[1]

Education is what survives when what has been learned has been forgotten.
B. F. Skinner
(*New Scientist,* 21 May 1964)

In the last chapter, we discussed how our cognitive model of student thinking helps us understand the importance of the ideas our students bring into the classroom. But cognition is complex. Students do not only bring ideas about how the physical world works into our classrooms. They also bring ideas about the nature of learning, the nature of science, and what it is they think they are expected to do in our class. In addition, they have their own motivations for success. We are often frustrated by the unspoken goal of many of our students to be "efficient"—to achieve a satisfactory grade with the least possible effort—often with a severe undetected penalty on how much they learn.

Most of my students expect that all they have to do to learn physics is read their textbooks and listen to lectures. Although some students who believe this don't actually carry out this minimal activity, even those who do often fail to make sense of physics in the way I want them to. This leads me to believe that reading textbooks and listening to lectures is a poor way of learning for most students. Sagredo objects, "This is clearly not universally true!" Remembering Principle 4, I concur. As physics teachers, most of us have had the experience of having a few "good" students in our lectures—students for whom listening to a lecture is an active process—a mental dialog between themselves and the teacher. Indeed, many of us have

[1]This chapter is based in part on a paper by Redish, Saul, and Steinberg [Redish 1998].

been that good student, and we remember lectures (at least some of them) as significant parts of our learning experience.[2]

A similar statement can be made about texts. I remember with pleasure working through some of my texts and lecture notes, reorganizing the material, filling in steps, and posing questions for myself to answer. Yet few of my students seem to know how to do this or even that this is what I expect them to do. This leads us to think about an additional observation.

Many of our students do not have appropriate mental models for what it means to learn physics.

This is a "meta" issue. People build schemas not only for content but also for how to learn and what actions are appropriate under what circumstances. Most of our students don't know what you and I mean by "doing" science or what we expect them to do. Unfortunately, the most common mental model for learning science in my classes seems to be:

- Write down every equation or law the teacher puts on the board that is also in the book.
- Memorize these, together with the list of formulas at the end of each chapter.
- Do enough homework and end-of-the-chapter problems to recognize which formula is to be applied to which problem.
- Pass the exam by selecting the correct formulas for the problems on the exam.
- Erase all information from your brain after the exam to make room for the next set of materials.

I call the bulleted list above "the dead leaves model." It's as if physics were a collection of equations on fallen leaves. One might hold $s = \frac{1}{2}gt^2$, another $\vec{F} = m\vec{a}$, and a third $F = -kx$. Each of these equations is considered to have equivalent weight, importance, and structure. The only thing one needs to do when solving a problem is to flip through one's collection of leaves until one finds the appropriate equation. I would much prefer to have my students see physics as a living tree.

A SECOND COGNITIVE LEVEL

The issues discussed in the introduction to this chapter seem to be at a different level of cognition than the more specific cognitive responses discussed in chapter 2. A number of cognitive researchers have identified a second level of cognition that resides "above" and controls the functioning of the level described in chapter 2 [Baddeley 1998] [Shallice 1988] [Anderson 1999]. Many of them refer to this as *executive function*—thinking processes that manage and control other thinking processes. In the context of instruction, three types of cognitive controls are particularly important: expectations, metacognition, and affect.

Each student, based on his or her own experiences, brings to the physics class a set of attitudes, beliefs, and assumptions about what sorts of things they will learn, what skills will be required, what they will be expected to do, and what kind of arguments and reasoning

[2] However, compare my discussion of one of my lecture experiences in the section of chapter 7 on "The Traditional Lecture."

they are allowed to use in the various environments found in a physics class. In addition, their view of the nature of scientific information affects how they interpret what they hear. I use the phrase *expectations* to cover this rich set of understandings that are particular to a given class. *Metacognition* refers to the self-referential part of cognition—thinking about thinking. Sometimes these responses are conscious ("Wait a minute. Those two statements can't be consistent."), but the term is also used to refer to the unconscious sense of confidence about thinking ("It just feels right."). Under *affect*, I lump together a variety of emotional responses including motivation, self-image, and emotion.

EXPECTATIONS: CONTROLLING COGNITION

Expectations affect what students listen to and what they ignore in the firehose of information provided during a typical course by professor, teaching assistant, laboratory, and text. They affect which activities students select in constructing their own knowledge base and in building their own understanding of the course material. The impact can be particularly strong when there is a large gap between what the students expect to do and what the instructor expects them to do.

Most physics instructors have expectation-related goals for their students, although we don't often articulate them. In our college and university physics courses for engineers, biologists, and other scientists, we try to get students to make connections, understand the limitations and conditions on the applicability of equations, build their physical intuition, bring their personal experience to bear on their problem solving, and see connections between classroom physics and the real world. Above all, we expect students to be *making sense* of what they are learning. I refer to learning goals like these—goals not listed in the course's syllabus or the textbook's table of contents—as part of the course's *hidden curriculum*.[3]

Expectations about learning

Students' expectations about how and what they will learn in science classes have been studied all across the curriculum. For pre-college, many studies have demonstrated that students often have misconceptions both about the nature of scientific knowledge and about what they should be doing in a science class.[4] Other studies indicate some of the critical items that make up the relevant elements of a student's system of expectations and beliefs. I focus here on studies at the college and secondary levels.

Two important large-scale studies that concern the general cognitive expectations of adult learners are those of Perry and Belenky et al. [Perry 1970] [Belenky 1986]. Perry tracked the attitudes of Harvard and Radcliffe students throughout their college careers. Belenky et al. tracked the views of women in a variety of social and economic circumstances. Both studies found evolution in the expectations of their subjects, especially in their attitudes about knowledge.[5] Both studies found their young adult subjects frequently starting in a *binary* or *received*

[3] The first use of this term that I know of is [Lin 1982].

[4] See, for example, [Carey 1989], [Linn 1991], and [Songer 1991].

[5] This brief summary is an oversimplification of a complex and sophisticated set of stages proposed in each study.

knowledge stage in which they expected everything to be true or false, good or evil, etc., and in which they expected to learn "the truth" from authorities. Both studies observed their subjects moving through a *relativist* or *subjective* stage (nothing is true or good, every view has equal value) to a *consciously constructivist* stage. In this last, most sophisticated stage, the subjects accepted that nothing can be perfectly known, and they accepted their own personal role in deciding what views were most likely to be productive and useful for them.

Although these studies both focused on areas other than science,[6] Sagredo and I both recognize a binary stage, in which students just want to be told the "right" answers, and a constructivist stage, in which students take charge of building their own understanding.[7] Consciously constructivist students carry out their own evaluation of an approach, equation, or result, and understand both the conditions of validity and the relation to fundamental physical principles. Students who want to become creative scientists will have to move from the binary to the constructivist stage at some point in their education.

An excellent introduction to the cognitive issues involved in student expectations is given by Reif and Larkin, who compare the intellectual domains of spontaneous cognitive activities that occur naturally in everyday life with those required for learning science [Reif 1991]. They point out that an important component of executive function is "deciding when enough is enough." They note that knowledge in the everyday domain is very much about *satisficing* rather than optimizing.[8] The kind of consistency, precision, and generality of principles typical of scientific knowledge is neither necessary nor common in people's everyday activities. Students often apply everyday-domain thinking when we want them to apply scientific-domain thinking.

The structure of student expectations: The Hammer variables

In order to get a handle on the complex issues of executive control and expectations, we need to begin defining specific characteristics so that we can talk about them and begin to think about ways to further them with instruction. In a series of interesting papers, David Hammer has begun this task [Hammer 1996a] [Hammer 1996b] [Hammer 1997]. In these papers, he identifies a number of parameters that arise from the expectations that a student brings into the physics class. Hammer's three variables are listed in Table 3.1.

I refer to these attitudes as *favorable* or *unfavorable,* since to make reasonable progress toward becoming a scientist or engineer, a student will find unfavorable attitudes limiting and will have to make a transition to the attitudes listed in the favorable column.

Sagredo complains, "I certainly expect my students to have the attitudes that you call favorable when they enter my class. If they didn't learn these attitudes in school, what can I do about it?" One of the problems, Sagredo, is that we often actually encourage unfavorable attitudes without really being aware of it. While working on his dissertation, Hammer did a case study with two students in algebra-based physics at Berkeley who were carefully matched

[6] Perry specifically excludes science as "the place where they *do* have answers."

[7] In my experience true relativism is rare, but not unheard of, among physics students.

[8] The term "satisfice" was introduced into economics and cognitive science by Herbert Simon, who won a Nobel Prize for the work. The point is that in real-world situations people do not optimize. It takes too much effort. Rather, they tend to seek an answer that is "good enough," that is, one that is both "satisfactory and suffices." This creates implications for the variational principles that economists construct.

Table 3.1 The "Hammer Variables" Describing Student Expectations [Hammer 1996a].

	Favorable	Unfavorable
Independence	takes responsibility for constructing own understanding	takes what is given by authorities (teacher, text) without evaluation
Coherence	believes physics needs to be considered as a connected, consistent framework	believes physics can be treated as unrelated facts or independent "pieces"
Concepts	stresses understanding of the underlying ideas and concepts	focuses on memorizing and using formulas without interpretation or "sense-making"

as to grade point average, SAT scores, etc., but who had decidedly different approaches to learning physics [Hammer 1989]. The first student tried to make sense of the material and integrate it with her intuitions. She didn't like what she called "theory" by which she said,

> It means formulas . . . let's use this formula because it has the right variable, instead of saying, OK, we want to know how fast the ball goes in this direction. . . . I'd rather know why for real.

The second student was not interested in making sense of what she was learning. For her, the physics was just the set of formulas and facts based on the authority of the instructor and text. Consistency or sense-making had little relevance.

> I look at all those formulas, say I have velocity, time, and acceleration, and I need to find distance, so maybe I would use a formula that would have those four things.

Student A was able to make sense of the material for the first few weeks. Soon, however, she became frustrated, finding it difficult to reconcile different parts of the formalism with each other and with her intuition. Eventually, she compromised her standards in order to succeed. Student B's failure to seek consistency or understanding did not hurt her in the course.

This small example indicates that we may inadvertently wind up encouraging students to hold unfavorable attitudes. After learning about these issues, I tried to change the way I taught in order to change this situation. How one might do this is discussed in chapter 4 on homework and testing and in chapter 5 on surveys and assessing our instruction. I used the Maryland Physics Expectations Survey (MPEX) we developed to test student expectations (described in chapter 5 and given on the Resource CD). Although at first I didn't get improvement, I learned that at least my grades were somewhat correlated with the results on my survey, whereas those of my colleagues were not. This can be taken in two ways: Either my survey is not measuring something we want students to learn, or our grades are not measuring those behaviors we want to encourage.

As we begin to develop a more complex view of what is going on in a physics class, what we want the students to get out of it, and what we value, we begin to realize that sometimes "the right answer" is not the only thing we should be looking for. A dramatic demonstration of student variability on attitudinal issues and how these issues play out in a classroom setting is given by Hammer's analysis of a discussion among a teacher and a group of high school students trying to decide whether a ball rolling on a level plane would keep moving at a constant speed [Hammer 1996a]. The students had been told Galileo's arguments that under ideal conditions it would do so.[9] I've numbered the lines in the discussion so that we can refer to them later.

1. Prior to this moment, the debate had mostly focused on the question of whether it is friction, gravity, or both that causes the ball to slow down. The students also debated whether it is appropriate to neglect friction or gravity, or both, and whether it is possible to neglect one without neglecting the other.
2. About 20 minutes into the debate, Ning argued that Galileo's ideal conditions would mean no forces on the ball, including no friction and no gravity; and, she claimed, "*if you don't put any force on it, it's going to stay still or go at constant speed.*" Bruce elaborated on Ning's statement, adding that there must be a force to make the ball move:
3. *Bruce:* If there is no gravity and no friction, and there is a force that's making it move, it's just going to go in a straight line at a constant speed. . . . What's making the ball move?
4. *Amelia* [over several other voices]: The forces behind it.
5. *Susan:* He [Galileo] said there was no force.
6. *Bruce:* If there's no force pulling it down, and no force slowing it down, it would just stay straight.
7. *Harry:* The ball wouldn't move.
8. *Jack:* There's no force that's making it go.
9. *Steve:* The force that's pushing it.
10. *Bruce:* The force that's pushing it will make it go.
11. *Jack:* Where'd that force come from, because you don't have any force.
12. *Steve:* No there is force, the force that's pushing it, but no other force that's slowing it down.
13. Many voices at once, unintelligible. Sean says he has an example.
14. *Teacher:* Sean, go ahead with your example.
15. *Sean:* If you're in outer space and you push something, it's not going to stop unless you stop it.
16. *Teacher:* If you're in outer space and you give something a push, so there's a place with no gravity—
17. *Sean:* No gravity, no friction.
18. *Teacher:* —it's not going to stop until you stop it. So Penny what do you think about that?

[9] Student names are pseudonyms.

19. *Penny:* But we talked about the ball on [a surface], but when we talk about space, it's nothing like space. So I was just saying that gravity will make it stop.

20. Amelia objected to Sean's example for another reason, saying that something moving in space will still stop.

21. *Amelia:* No. Maybe there's no gravity and no air there, but there are other kinds of gases that will stop it.

22. *Teacher:* But those are other, those are outside things.

23. *Amelia:* The outside friction should stop it.

24. *Bruce:* That's not, that makes it an un-ideal state.

25. *Scott:* Space is a vacuum. Like a vacuum there's no—

26. *Amelia:* There are other kinds of gases.

27. [Several voices, unintelligible.]

28. *Harry:* We're talking about ideal space. (students laugh)

29. I intervened at this point to steer the discussion away from the question of whether there are gases in space and toward the question of whether there is a "force that's moving" the ball.

30. *Teacher:* . . . So how can one side say there are no forces on it, and the other side say there is a force that's moving it.

31. *Bruce:* Well there was an initial force.

32. *Susan:* There's an initial force that makes it start, giving it the energy to move.

In analyzing this discussion, Hammer identifies half a dozen perspectives that could be used to evaluate the students' responses. I want to focus on four.

- *Content answer:* Does the student have the correct answer?
- *Reasoning:* Does the student display a common naïve conception? Is it related to a reasoning primitive?
- *Coherence:* Does the student understand that scientific laws are developed to unify a wide variety of circumstances and that science should be consistent?
- *Understanding idealizations:* Can the student see the relevance of idealized or limiting conditions?

In the dialog, Ning gave the correct answer (line 2) but did not participate in defending it. The discussion revealed that many of the students had the common naïve conception represented by the facet "motion is caused by force" (lines 3, 8, 11, 12). Almost all of the discussion was by claim and counterclaim without citing reasoning or evidence. The discussion in lines 15–19 shows a distinction between Sean, who is trying to make a link between two rather different physical situations, and Penny, who wants to keep them separate. This can be interpreted as a difference in their understanding of the need for coherence in science. Sean's claim in line 15 tried to take the analysis to an idealized situation, without gravity or friction. Amelia (lines 23 and 26) did not appear to be comfortable in thinking about the simplified example.

In other examples cited by Hammer, students gave the correct answer to a problem, but argued its validity by citing the text or teacher and being unwilling to think about the issue for themselves.

These examples illustrate the complexity of our hidden curriculum and show how we can begin to think both about what the student is bringing in to our classes and what the student can gain from our classes in a more sophisticated way than just "are they giving the right or wrong answers."

Connecting to the real world

Although physicists believe they are learning about the real world when they study physics, the context dependence of cognitive responses (see chapter 2) opens another possible gap between faculty and students. Most students seem to believe that physics is related to the real world in principle, but a significant fraction also believe that what they are learning in a physics class has little or no relevance to their personal experience. This can cause problems that are both serious and surprising.

Even if our students develop strong concepts related to real-world meanings, the strong context dependence of the cognitive response makes it particularly easy for students to restrict their learning in physics classes to the context of a physics class. This seems unnatural to Sagredo. "Practically every problem I assign for homework or do on the board involves some real-world physical context." True, Sagredo. But that doesn't mean that students will easily or naturally make the connections that you do.

When an instructor produces a demonstration that has been "cleaned" of distracting elements such as friction and air resistance, the instructor may see it as displaying a general physical law that is present in the everyday world but that lies "hidden" beneath distracting factors. The student, on the other hand, may believe that the complex apparatus is *required* to produce the phenomenon and that it does not occur naturally in the everyday world, or is irrelevant to it. A failure to make a link to experience can lead to problems not just because physics instructors want students to make strong connections between their real-life experiences and what they learn in the classroom, but because learning tends to be more effective and robust when linked to real and personal experiences.

Even worse, students' failure to connect their personal experience to what is happening in their physics class can put up barriers to understanding that grow increasingly impenetrable. As discussed in chapter 2, multiple representations are used in physics in order to code knowledge in a variety of interlocking ways. A critical element in all of them is the map to the physical system. An essential part of solving a problem is understanding what the real-world version of the problem is, what's important in that situation, and how it maps onto physical principles and equations. If students don't understand that part of the process, they can have great difficulty in seeing the physics as a way to make sense of the physical world.[10]

[10] The Physics Education Group at the University of Massachusetts—Amherst has done interesting research using problem posing as a technique to help students develop these skills [Mestre 1991]. See also the variety of problems discussed in chapter 4.

A shepherd has 125 sheep and 5 dogs. How old is the shepherd?

Figure 3.1 A word problem for middle school math students.

A classic word problem that illustrates this difficulty is shown in Figure 3.1. Although this problem is patently absurd and cannot be answered, some middle school students will struggle to find an answer (Expectation: "The teacher wouldn't give me a problem that has no solution.") and will come up with an answer of 25. ("There are only two numbers to work with: 5 and 125. Adding, multiplying, and subtracting them doesn't give something that could be an age. Only dividing gives a plausible number.")

Another example comes from the mathematics exam given by the National Assessment of Educational Progress (NAEP). A national sample of 45,000 13-year-olds was given the problem shown in Figure 3.2 [Carpenter 1983]. Although 70% of the students who worked the problem carried out the long division correctly, only 23% gave the correct answer—32. The answer "31 remainder 12" was given by 29%, and the answer 31 was given by another 18% of those doing the problem. Thus, nearly half of the students who were able to carry out the formal manipulations correctly failed to perform the last simple step required by the problem: to think about what the answer meant in terms of a real-world situation. (Expectation: "The mathematical manipulation is what's important and what is being tested.")

In these two examples, students are making somewhat different errors. In the shepherd problem they *are* using some real-world information—what ages are plausible as answers; but they are not asking how the numbers they are given could relate to the answer. They are not making sense of the problem. In the soldiers and buses problem, students are *not* using their real-world knowledge that you cannot rent a fraction of a bus. In both cases, students who make these errors focus on the mathematical manipulations and fail to "make sense" of the problem in real-world terms.

The same problems occur frequently in introductory physics. In my experience with introductory college physics, more than half of the students do not spontaneously connect what they learn in their physics class to their everyday experiences—either by bringing their everyday experiences into their physics classes or by seeing the physics they are learning in the outside world. Two anecdotal examples show how this plays out in a college physics class.

A student in my algebra-based physics class missed a midsemester exam due to an illness, and I agreed to give her a makeup. One of the problems on the exam was the following. "A high jumper jumps so his center of gravity rises 4 feet before he falls back to the ground. With what speed did he leave the ground?" This is a typical projectile problem. My

An army bus holds 36 soldiers. If 1128 soldiers are being bused to their training site, how many buses are needed?

Figure 3.2 A problem for the NAEP math exam for middle school students.

student knew the formula and punched the numbers into her calculator. When she handed in her test and I looked over her answers, she had come up with the answer 7840 feet/second. (Can you guess what she had done wrong on her calculator?) I asked her whether her answer to that problem had bothered her. She shrugged and said, "That's what the formula gave me." She saw absolutely no need to check her answer against her experience—and incidentally, it had never entered her mind that she might have misremembered the formula, incorrectly recalled the value of a parameter, or made an error in pressing the calculator keys. This overconfidence in their memory and processing is a symptom I have seen in very many students. They assume anything they remember must be correct.

A second example occurred in my engineering (calculus-based) physics class. For many years now, I have been requiring estimation (Fermi-type) problems in my classes.[11] Almost every homework assignment has one, and every exam is guaranteed to have one. One of my students came into my office hours and complained that this wasn't fair. "I don't know how big these things are," she scowled. "Well," I said. "How about a foot? Do you know how big a foot is?" "I have no idea," she replied. Assuming that she was overstating her case to make her point, I said, "How about making a guess? Show me how far up from the floor a foot would be." She placed her hand at about waist level. "And how tall are you?" I asked. She thought for a second, said "Oh" and lowered her hand somewhat. She thought again and lowered her hand again—to about the right height above the ground. She looked at her hand—and at her foot a few inches away and remarked with great (and what appeared to be genuine) surprise, "Oh! Does it have anything to do with a person's foot?"

Since these real-world connections are critically important in developing an understanding of how physics helps us to make sense of our everyday experiences,[12] I specify a fourth learning goal.

> Goal 4: Reality Link—Our students should connect the physics they are learning with their experiences in the physical world.

To what extent does a traditional course help our students reach this goal? The simplest way to find out is to ask them.[13] In our study of student expectations in a calculus-based physics class for engineers [Redish 1998], using the MPEX survey[14] we found that student expectations of the connection between physics and the real world typically tended to deteriorate as a result of the first semester of instruction.

The four items of the MPEX reality cluster are shown in Table 3.2. They ask whether students expect to/have needed to[15] make the link to their outside experiences for the class and whether students expect to/have found that what they learn in physics can be seen in

[11] For examples of these types of problems, see chapter 4 and the sample problems on the Resource CD.

[12] This is especially true for our service students in engineering and biology.

[13] This method is not very accurate since students often do not reflect and do not necessarily know how they think. A better approach is to watch them solving problems alone or in a group using think-aloud protocols [Ericson 1993].

[14] See chapter 5 for a detailed discussion of the MPEX. The full survey and instructions on its use are contained in the Action Research Kit on the Resource CD.

[15] The alternate forms are for the pre- and post-class surveys.

TABLE 3.2 Results on the MPEX Reality Link Cluster Items

MPEX Item	Favorable Pre	Unfavorable Pre	Favorable Post	Unfavorable Post
Physical laws have little relation to what I experience in the real world. (−)	84%	5%	87%	2%
To understand physics, I sometimes think about my personal experiences and relate them to the topic being analyzed. (+)	59%	11%	54%	22%
Physics is related to the real world, and it sometimes helps to think about the connection, but it is rarely essential for what I have to do in this course. (−)	73%	9%	61%	19%
Learning physics helps me understand situations in my everyday life. (+)	72%	10%	51%	18%

their real-world experiences. Both issues are addressed in two statements, one positive and one negative. The student's response is considered to be *favorable* if she sees the need for a connection and *unfavorable* if she does not. The polarity of the favorable result is indicated after the item by a (+) when the favorable result is *agree* and by a (−) when the favorable result is *disagree.* The students are asked to report on a five-point scale (strongly agree, agree, neutral, disagree, strongly disagree), but for a favorable/unfavorable analysis, we ignore whether or not there is a "strongly." The responses come from pre- and post-surveys given in the first semester of an engineering physics class. The class was calculus-based and covered mostly Newtonian mechanics. The results are shown for $N = 111$ students (matched, i.e., who completed both pre- and post-surveys).[16]

The results are discouraging, especially on the last two items. I tried to help my students make the connection by giving some estimation problems, but that was clearly insufficient. Similar results have been found with other faculty teaching this class at Maryland and at many other colleges and universities [Redish 1998].

There has been little published work on how to help students develop a strong reality link. In my experience, regular essay questions asking the students to relate the physics to their experience and regular estimation questions (being sure to include both on every exam so that students take them seriously) only help a little bit. Even in lessons where physicists see real-world implications immediately, students rarely make the connections spontaneously

[16]A total of 158 students completed the class.

if not led to them. I expect this goal will only be achieved by a thorough interweaving of the physics with explicit connections to the students' experience.[17] Further research and development on this issue would be most welcome.

METACOGNITION: THINKING ABOUT THINKING

The transcript from David Hammer's high school class in our earlier discussion shows that different students access different kinds of reasoning in their discussion of a physics problem. This variety arises from students having different expectations about the nature of science and what it means to learn science. Unfortunately, many of these expectations are inappropriate for learning science. They may be learned in school, from movies and TV, or from reading science fiction books.[18] When students have the wrong expectations about what they are supposed to do in a class, those expectations can serve as a filter, causing them to ignore even explicit instructions given by the instructor.

In part, the approaches to learning physics that students bring into our classes arise from a misunderstanding of the nature of scientific knowledge and how one has to learn it. As pointed out so clearly by diSessa and discussed in chapter 2, for most ordinary people (even for some of our best students[19]) knowledge of the world comes in "pieces" about how particular situations work [diSessa 1993] [diSessa 1988]. As pointed out by Reif and Larkin [Reif 1991], building a consistent and economical set of principles—at the cost of requiring long and indirect explanations of many phenomena—is not the way most people create their models of the physical world in their everyday lives. It seems that people tend to look for quick and direct explanations. The complex consistent and parsimonious net of links built by science is not a natural type of mental construction for most people. It has to be learned.

The key element in the mental model I want my students to use in learning physics appears to me to be *reflection*—thinking about their own thinking. This includes a variety of activities, including evaluating their ideas, checking them against experience, thinking about consistency, deciding what's fundamental that they need to keep and what is peripheral and easily reconstructed, considering what other ideas might be possible, and so on. My experience with students in introductory classes—even advanced students[20]—is that they rarely expect to think about their knowledge in these ways. Students often come to my office hours for help with problems. I always ask them to show me what they have tried so far and proceed to offer help via questions. They frequently have an error close to the start of their analysis—in a principle or equation that they bring up from their memory. As I lead them to implausible and unlikely results through my questioning they become troubled, but they are much more likely to try to justify a ridiculous result by difficult and inconvenient contorted reasoning than by asking if one of their assumptions might be wrong.

[17] Preliminary results with a more synergistic approach appear quite favorable [Redish 2001].

[18] Some science fiction books, especially those written by scientists (such as David Brin, Gregory Benford, or John Kramer), have excellent descriptions of the way science develops its knowledge.

[19] Recall that in [diSessa 1993] the subjects studied were MIT freshmen.

[20] Many of the students in my algebra-based physics classes are upper division students who have previously taken many science classes in chemistry and biology.

From our cognitive model we understand that to create new, coherent, and well-structured mental models, students need to go through a number of well-designed activities addressing the issue to be learned, to repeat them, and to reflect on them. Similar principles hold for metacognition—thinking that reflects on the thinking process itself. I add another learning goal to the list developed in chapter 2.

> Goal 5: Metalearning—Our students should develop a good understanding of what it means to learn science and what they need to do to learn it. In particular, they need to learn to evaluate and structure their knowledge.

This is not a trivial goal and it does not happen automatically for most students as they work to learn physics content.

> *Redish's second teaching commandment:* In order for most students to learn how to learn and think about physics, they have to be provided with explicit instruction that allows them to explore and develop more sophisticated schemas for learning.

"Hold on!" Sagredo complains. "I never have time enough to teach all the content I'm supposed to teach. How can I find time to give them lessons in how to learn?" I sympathize, Sagredo. But in fact, the problem is not as bad as it looks. If we are teaching them to learn, we have to be teaching them to learn *something*. That something can easily be the appropriate physics content. Some introductory discussion, lessons designed to encourage particular activities, and reflections analyzing what they've done should help substantially. One of the few well-documented approaches to explicitly teaching and improving students' metacognition is the work of Alan Schoenfeld.

Instructional techniques for improving metacognition

Alan Schoenfeld, in a problem-solving college math class, developed a group-problem-solving method that focused on helping students strengthen their judgment and control of their own thinking. The class was small enough (on the order or fewer than 25 students) that he could use a guided cooperative group-problem-solving approach.[21]

In his observations of the class's behavior, Schoenfeld found that his students often wasted a lot of time in following unproductive approaches through a lack of metacognitive activity. The students quickly jumped on the first idea that came to their minds and then proceeded to "churn" through extensive manipulations, frequently losing track of what they were doing and rarely evaluating whether their approach was productive.

Schoenfeld developed an instructional method to help students become more metacognitively aware. The key was the mantra of metacognitive questions posted on the wall shown in Figure 3.3. His comments on how this worked are worth repeating.

> Students' decision-making processes are usually covert and receive little attention. When students fail to solve a problem, it may be hard to convince them that their failures may be due to bad

[21] See chapter 8 for a discussion of a method of this type employed in physics to help develop students' conceptual development and problem-solving skills.

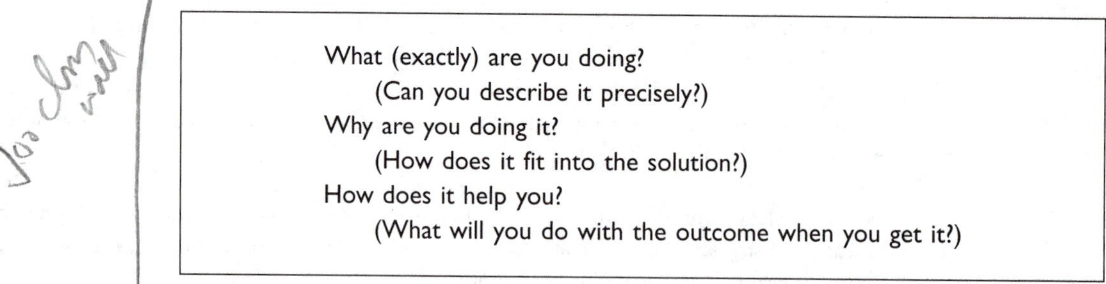

Figure 3.3 Schoenfeld's questions for helping students learn to focus on metacognitive issues.

> decision-making rather than to a lack of knowledge. The instructor had the right to stop students at any time while they were working on the problems and to ask them to answer the three questions on [Figure 3.4]. At the beginning of the course the students were unable to answer the questions, and they were embarrassed by that fact. They began to discuss the questions in order to protect themselves against further embarrassment. By the middle of the term, asking the questions of themselves (not formally, of course) had become habitual behavior for some of the students. [Schoenfeld 1985]

Schoenfeld not only implemented a focus on metacognition and control in the group activity, but he modeled it in his approach to modeling solutions for the class as a whole. His description outlines the process in detail.

> When the class convened as a whole to work problems (40–50% of class time), I served as orchestrator of the students' suggestions. My role was not to lead the students to a predetermined solution, . . . my task was to role model competent control behavior—to raise the questions and model the decision-making processes that would help them to make the most of what they know. Discussions started with "What do you think we should do?" to which some student usually

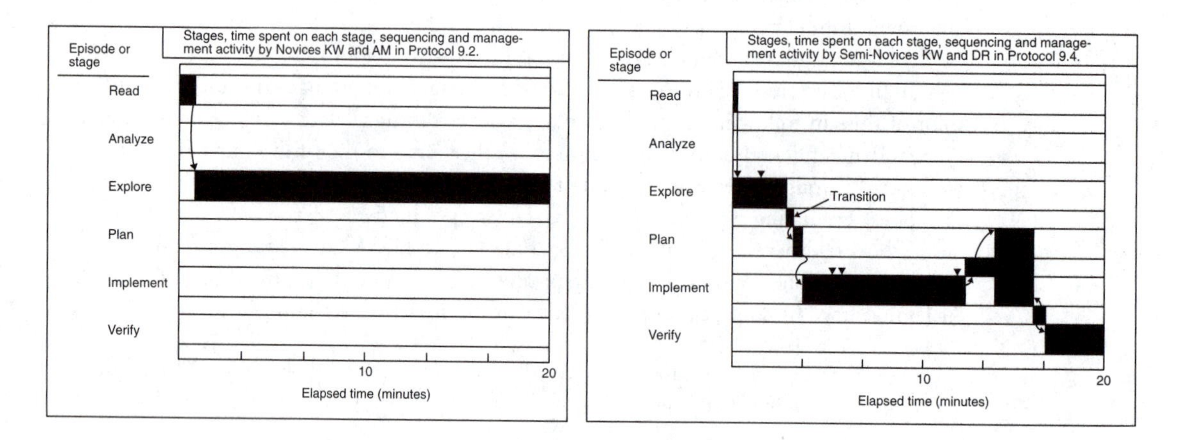

Figure 3.4 Sample plots of student activities in solving math problems in Alan Schoenfeld's metacognitive math class. Small triangles mark metacognitive statements [Schoenfeld 1985].

> suggested "Let's do X." Often the suggestion came too rapidly, indicating that the student had not adequately thought through what the problem called for or how the suggestion might be useful. The class was then asked, "Are you all sure you understand the problem, before we proceed with X?" A negative response from some students would result in our taking a closer look at the problem. After doing so, we returned to X as a possible solution approach. Did X still seem reasonable? Not infrequently the answer was "no." When it was, this provided the opportunity to remind students about the importance of making sure that one has understood a problem before jumping into its solution. . . . After a few minutes of working on the problem—whether or not we were on a track that would lead to a solution—the process would be halted for an assessment of how things were going. The class was asked "We've been doing this for 5 minutes. Is it useful, or should we switch to something else? (and why?)" Depending on the evaluation, we might or might not decide to continue in that direction: we might decide to give it a few more minutes before trying something else. Once we had arrived at a solution, I did a postmortem on the solution. The purpose of that discussion was to summarize what the class had done and to point out where it could have done something more efficiently, or perhaps to show how an idea that the class had given up on could have been exploited to solve the problems. . . . The same problem was often solved three or four different ways before we were done with it. [Schoenfeld 1985]

By the end of the class, Schoenfeld found that the students were spending a much larger fraction of their time in planning and evaluation and that their "metacognitive events" (statements like: "I don't understand this" or "That doesn't seem right") more often led to their jumping into planning or checking mode than it did at the beginning of the class. This is illustrated in Figure 3.4.

AFFECT: MOTIVATION, SELF-IMAGE, AND EMOTION

It is patently clear to most university physics instructors that motivation, how students feel about the class, and how the students feel about themselves, play a critical role in how students respond to instruction and how well they learn. The issues of feeling, emotion, and mood are summarized by the term *affect* or *affection* in psychology. These issues have been discussed extensively in the educational literature, [Graham 1996] [Stipek 1996], but I do not attempt to review this literature here as the interaction between affect and cognition is extremely complex and it is difficult to provide precise guidance. This is not to say these issues are not of great importance. I therefore make a few comments, but refer the reader to the literature cited above for details.

Motivation

Motivation can be a major factor in distinguishing students who will make the effort to learn and those who will not. We encounter a variety of motivations.

- *Internally motivated*—Some students who come to our classes are self-motivated by an interest in physics and a desire for learning.
- *Externally motivated*—Some students have no internal interest in physics but are strongly motivated to get a good grade because our class is a hoop that must be jumped through for them to get into a program for which they are motivated.

- *Weakly motivated*—These students are taking physics because it is a requirement, but they are concerned only about passing, not getting a good grade.
- *Negatively motivated*—Some students are motivated to fail—for example, in order to demonstrate to a controlling parent or mentor that they are not suited to be an engineer or a doctor.

Those in the first group are a physics instructor's delight. Whatever you give them they make the most of. We can work with those in the second group by controlling the learning environments we set up and making clear what will be evaluated on exams. (See examples in chapter 4.) I can rarely do anything with the last group. Their goals in the class are distinctly different from mine.

Finding ways to motivate your students to want to learn physics can be an extremely effective lever to improve the success of your teaching. Unfortunately, this is easier said than done and is where much of the "art" in teaching comes in. It is easy to mistake student happiness for student motivation. Making your lecture "entertaining" does not necessarily increase students' motivation for learning. Indeed, it can set up the expectation in their minds that associates your lecture with a TV program where they don't have to think.

Providing connections to their chosen career sometimes helps. I evolve my estimation problems into design problems in my engineering physics class and create problems with a medical and biological context for my algebra-based students. I hope this helps them see the relevance of physics to a profession toward which they should, in principle, be motivated. (Interviews with a small number of volunteers—usually the better students—suggest that at least this group is making the connection [Lippmann 2001].)

Motivation is perhaps the primary place where the teacher in fact makes a significant difference. A teacher with the empathy and charisma to motivate the students can create substantially more intellectual engagement than one who reads from the book and does not take the time to interact with the students. Perhaps the most critical element in creating motivation is showing your students that you are interested in them, you want them to succeed, and you believe that they can do it.

Self-image

Sagredo is a bit skeptical about the issue of students' self-image. He feels that the education community pushes "helping students feel good about themselves," sometimes to the detriment of serious critical self-analysis and learning, at least if the letters to the editor published in newspapers are to be believed. In my experience with university-level physics students, this issue cuts two ways. Some students are supremely overconfident, while others think that they cannot possibly understand physics. Both groups are difficult to deal with.

In our small-group-sessions, we often use the Tutorial materials developed at the University of Washington. These lessons are research-based group-learning worksheets (see chapter 8 for a detailed description) and use a cognitive conflict model. As a result, students who are used to being right often feel the Tutorials are trivial and therefore useless—even when they are consistently getting the wrong answers. When I am facilitating in one of these sessions, I see this as a terrific learning opportunity. I circulate through the class, asking what they got on the tricky questions. When I find a group that has been overwhelmed by an overconfident student with a wrong answer, I say, "Now remember: Physics is not a democracy and physics is not determined by charisma. You can't tell who's right by who says it the most

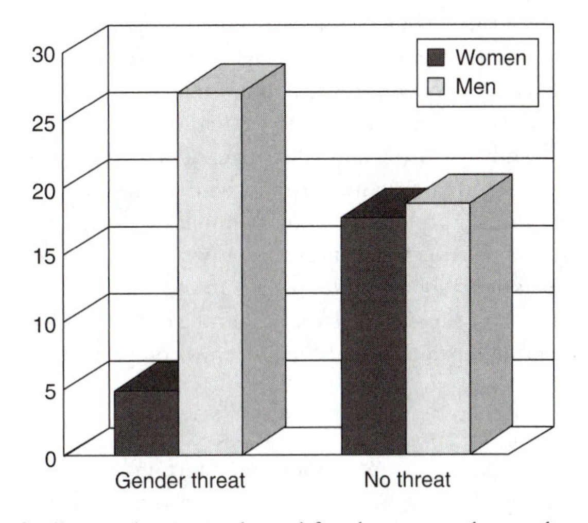

Figure 3.5 Scores of college sophomore males and females on a math test when a comment is made that the test "tends to separate genders" and when no such comment is made [Steele 1997].

forcefully or by what most people think. It has to make sense and it has to be consistent. Perhaps you want to go back and think that question out again." The result is almost always that someone else in the group who had previously been intimidated into silence can bring everyone to the correct result. This sends a really useful message—both for the overconfident student and for the other members of the group.

On the other side, I have had experience with students who were absolutely convinced that they were incapable of learning physics. In one case, I had a student in algebra-based physics who was convinced "she couldn't do this stuff" and told me so repeatedly. However, I often watched her vigorously argue difficult issues in Tutorials with another student who was supremely confident of her ability and answers. My underconfident student was almost always right, and my overconfident student almost always wrong.

Despite her success in Tutorials, this student did not change her overall self-evaluation of her ability and she did poorly on exams. In other cases, I was able to help students who were good in other classes but who, perhaps because of bad experiences in high school, were convinced that they "couldn't do physics." All these cases are best treated carefully and individually, using all the empathy and understanding you can bring to bear. Unfortunately, in many college and university situations, the pressure of time and numbers makes it difficult, if not impossible, to allow you to offer the individualized responses needed.

There has been some research on the topic of math anxiety or "math phobia." (See, for example, [Tobias 1995].) I do not know of comparable work on "science phobia." There has also been some extremely important work on the implications of social stereotypes on self-image and performance. Stanford sociologist Claude Steele explored the implications of raising the link in a student's mind to gender or race in conjunction with a mathematics test [Steele 1997]. College sophomores who had committed themselves to a math major or minor were given a test somewhat above their level. One group was told that the test was "just a trial" and that the researchers "wanted to see how they would do." A second group was told up front that the test "showed gender differences." (The sign of the difference was not specified.) The results, shown in Figure 3.5, were dramatic. In the group given the test without any comment about gender, males and females scored approximately the same. In the group with the comment about gender, referred to as a *gender threat* by Steele,[22] females scored significantly worse (by more than a factor of 3!), and males scored somewhat better (about 50%).

The implication appears to be that stereotypes (males are better in math) pervade our culture in a profound way, with implications that we tend to be unaware of and are insensitive to. This certainly suggests that we should be extremely cautious about making *any* comments at all about gender or race to our classes. For researchers, it suggests that in doing interviews or surveys, questions about the respondent's gender, race, or other social factors should be given separately after the testing is complete.

Emotion

"I'm a physicist, not a song-and-dance-man!" Sagredo complains, echoing Star Trek's Dr. McCoy. Perhaps, Sagredo, but making your students feel good about your class can have an influence on their learning. For one, if they hate your lectures and don't come to class, they won't be able to learn anything from them.[23] On the other hand, if you fill your lecture with jokes, films, and cartoons, they are unlikely to take them seriously.

The best thing you can do to make students "feel good" about your class is to make it worthwhile, at an appropriate level, and fair. Students like to feel that they are learning something valuable and that they can get a "good" grade (this may have different meanings for different students) without having to work so hard that their other classes (and their social lives) suffer. Getting students to learn a lot from our classes is a process of negotiation. From my point of view as a teacher, I want them to work hard, but from their point of view as a student, they don't want to work hard without a clear payoff. In physics, learning can be frustrating and nonlinear. Often you have to work for a long time without feeling that you are making much progress. Then, suddenly, everything falls into place and it all makes sense. But until the "click," you can't be sure how much time you need to "get it" and it's difficult to plan. Students first have to learn what understanding the physics feels like and be slowly drawn into working hard enough to learn harder and harder topics.

But entertainment and "song-and-dance" don't have to be shunned, Sagredo. In our context, it can mean little physics jokes, personalized stories, and dramatic demonstrations. (But see the discussion of demonstrations in chapter 7.) All of these can be effective—or not. Jokes should be relevant, not off-color, and not derogatory to groups or individuals. Personalized stories should be relevant to the physics involved and have some point that will make sense to a novice. They shouldn't occupy so much of the time that students begin to feel you're not offering them enough physics. Demonstrations can be the best but are also dangerous. As explained in chapter 7, demonstrations can be entertaining but misleading. Students often don't see what you think they are seeing. A careful and involving class discussion, both before and after the demonstration, is usually needed.

[22] Note that the "threat" is implicit. There was no statement as to which group was expected to do better, and there were no consequences for the students no matter how they scored.

[23] Students tend to learn little even from lectures they attend unless special tools are used. See chapter 7.

CHAPTER 4

Extending Our Assessments: Homework and Testing

If I had to reduce all of educational psychology
to just one principle, I would say this:
The most important single factor influencing learning
is what the student already knows.
Ascertain this and teach him accordingly.
D. P. Ausubel [Ausubel 1978]

The two-level cognitive model we develop in chapters 2 and 3 has implications for the way we evaluate our students and our instruction. First, our understanding of the context dependence of the cognitive response leads us to appreciate the importance of students developing functionality and coherence in their knowledge, not just learning isolated facts or memorizing a small set of problem solutions. Second, our understanding of the associative character of the cognitive response leads us to appreciate the importance of building a good conceptual base and helping students link their conceptual understanding to their problem-solving and analysis skills. Third, our understanding of student expectations and the stability of well-established mental models leads us to appreciate that changing what our students value and what they choose to do to learn is going to require a negotiation between instructor and students.

This negotiation has two important components: (1) We need to offer our students activities that promote the kind of thinking we want to encourage (seeking coherence, sense-making, effective analysis, creativity and so on), and (2) we need to set up mechanisms that permit two-way feedback.

What students actually "do" in our classes usually consists of a variety of activities including both in-class activities, such as listening to a lecture or doing a lab, and out-of-class activities, such as reading the text or doing homework.[1]

[1] In the education literature, activities that are done by all students together in a class are referred to as *synchronous*. Activities that are done by students out of class independently at their own time and pace are called *asynchronous*.

In a negotiation, there must be feedback in both directions. We, the instructors, need to know what our students are doing and how much they have learned. Students in our classes need to know when they are meeting the goals we have for them and how to modify their activity when things aren't working.

In this chapter, I discuss the role of homework and testing in our classes. (Other components of the class are discussed in chapters 7–9.) These activities play critical roles in helping our students understand what it means to learn and do physics. I begin with a general discussion of the goals of assessment and evaluation. Next I specifically discuss homework and testing and present some of the methods I have found useful. I then review some of the different kinds of questions we can use both in homework and on exams to help our students understand what it is we want them to get out of our course and to help us understand the extent to which they have done it.

ASSESSMENT AND EVALUATION

If we are interested in probing how well our instruction is working, we can think about the answer along two axes, depending on our answer to the questions "when?" and "what?"

- *When?* Are we asking while the class is still in session or at the end?
- *What?* Are we asking about a student or about the class as a whole?

Probes of the success or failure of our instruction that occur during the class can help both the student and the instructor correct things that are going wrong and identify problems that still have a chance to be fixed. Such probes are called *formative.* Probes that occur at the end of a class and that serve to provide a final certification of student success are called *summative.* There is also a difference in whether we are probing individual students or our instruction as a whole. I refer to probes of an individual student's learning as *assessment* and to probes of our instruction as a whole as *evaluation.*

In this chapter, I focus on looking at assessment through the lens of the question, "What has a particular student learned?" In the next chapter, I consider evaluation and the question, "How effective was my instruction for the group of students in my class?" This separation is somewhat artificial, since when we assess our individual students' learning it tells us something about how successful our instruction has been overall. When we try to evaluate our instruction overall, the only way we can do it is by probing individual students. We will see, however, that some tools that are useful for evaluation of an overall result may not do an adequate job of assessing an individual's learning.

To get feedback on how our class is doing and to measure the success of our interventions, we must first decide what we mean by "success." The discussions in chapters 2 and 3 show that this can be a complex issue. Our choice of goals plays an important role in determining our approach to assessment. What we mean by success is, in turn, determined by our model of student understanding and learning.

To evaluate a student's learning, we have to take account of a number of factors, including conceptual learning, problem-solving skills, coherence and appropriateness of their patterns of association/mental models, and so forth. We must recall that a student may "have" an item of knowledge, that is, be able to recall it in response to a narrow range of cues, but be unable to recall and apply it in a wide range of appropriate circumstances. As a consequence, our probes need to be varied and diverse. Furthermore, we need to watch out for "wishful thinking" such as "filling in the blanks." Often, a student will offer a statement or an equation, and we will assume that they have come to that result in the same way that we would have. Unfortunately, that is often not the case.

GIVING FEEDBACK TO YOUR STUDENTS

As we learned in chapters 2 and 3, students can have ways of looking at and analyzing physics that are quite different from those expected by the teacher. Significant conceptual difficulties may be buried in a student's "wrong" answer to a homework or exam problem, and the student may not have the knowledge or self-evaluation skills to disentangle them. Real feedback on their thinking is immensely valuable to them. Unfortunately, this is one of the first things to be cut when financial or personnel constraints get tight and classes get large. Homework grading disappears in favor of only assigning problems for which the answer is given in the book. Detailed solutions can help somewhat, but since there are often many ways to approach a particular problem, they do not always help students debug their own thinking.

> *Redish's third teaching commandment:* One of the most useful aids you can give your students is detailed feedback on their thinking—in an environment in which they will take note of and make use of it.

HOMEWORK

Sagredo says that he always assigns homework in his physics classes, since it's in doing the problems that the students "really learn the physics." Here's a place where he and I strongly agree. In my experience, far too many students have the expectation that physics is a set of simple "facts" and a list of equations, which, if memorized, should permit them to get an A in my class. Far too few are able to solve the complex kinds of problems that require them to think deeply about the physics and to build a strong conceptual understanding and an appropriate and well-organized knowledge structure. Homework can be of immense value in helping students learn physics, but the wrong kind of homework can also send the wrong message, confirming their view that "facts and equations are enough."

Unfortunately, sometimes despite our best efforts, the wrong message still gets through. End-of-chapter problems given in most introductory texts come in a variety of formats—questions (often conceptual), exercises (straightforward plug-and-chug), and problems (more complex reasoning required). Sagredo and I have both assigned a mix of problems for many years. But recently, I have come to suspect that for many of our students this is still doing them a disservice. A significant fraction of my students appear to think (or hope) that what my course is really about is straightforward plug-and-chug. This is a pretty standard wishful thinking, since these are the problems they can comfortably do. If they write something arbitrary on the questions, get all the exercises right, and write down enough to get partial credit on the problems, they are satisfied that they have gotten all that is necessary out of the assignment. In fact, these students largely miss the main point. The questions are supposed to get them to think deeply about the concepts, and the problems are supposed to help them build a good understanding of how to work with physics. Given the "escape hatch" of the exercises, some students make little effort to get any farther.

As a somewhat draconian attempt to close this loophole, I have stopped giving students exercises as part of their homework. Instead, I reduce the total number of problems and give a smaller number of harder problems, including essay questions and context-rich problems. (See the explanation of these problems below.) One of my better students complained about this. He said he needed the exercises to gain familiarity with the material. I responded that he was perfectly welcome to do them and that many had answers in the back of the book so he could check his work—but that success in this class means much more than plug-and-chug problems and that he would be evaluated on what I wanted him to learn in the end.

Even when we give complex problems, we tend to "degrade" them in response to student pressure. As described in chapter 2 in the section on multiple representations, a major difficulty in solving a realistic physics problem is the extraction of a solvable physics problem from a complex real-world situation. When we assign variable names to the quantities in the problem situation, when we only mention quantities that are relevant and give no extraneous information, when we set up the problem so that all choices of what's important and what's not are explicitly indicated, we steal from the student the opportunity to learn a whole range of essential problem-solving skills.

When students find some tasks difficult, we have a tendency to give in to their pressure and modify the problems so that there is no need for them to develop those skills. A crucial example is the ability to manipulate and interpret equations containing many symbols. Sagredo is aware that students have this difficulty and is disturbed by it. "Why should it matter? A symbol just stands for a number anyway. They should be able to use equations either way." I agree, Sagredo, but recall that experience and familiarity with representations and manipulations make a big difference in the ability to recognize appropriate actions.[2] I once looked carefully at a couple of standard introductory calculus texts. To my surprise I found that very few equations anywhere in the text contained more than one symbol. Those that did followed a strict convention: x, y, z, and t were variables, while a, b, and c were constants. If we want students to learn the skill of working with multisymbol equations, we need to provide examples that will make them use it. Problems in which numbers are given right away allow the students to avoid having to learn the skill of interpreting equations with multiple symbols.

I find that weekly homework using a variety of challenging problems (especially if some are marked "from an hour exam" or "from a final exam") can be effective in helping to reorient students' ideas of what they need to do to learn physics.

In addition to providing a rich venue for student activity, homework also plays a critical role in providing students with formative feedback. In many large classes, we cannot give quizzes and examinations frequently enough to provide students enough feedback to modify a learning approach they may have developed in high school or in other science classes. As I discussed in chapter 3, student expectations play a major role in what they do. Even if you give them good feedback, there is no guarantee that they will use it. If homework problems are graded with comments but handed back weeks after they were done, the students might

[2]Recall, for example, the Wason card (K2A7) example discussed in chapter 2.

not be able to reconstruct their state of mind when they did the problems. Any difficulty they have interpreting written feedback is likely to result in their ignoring it.

GETTING FEEDBACK FROM YOUR STUDENTS

The Ausubel quote in the epigraph, our cognitive model, and my experience all agree: Feedback needs to work both ways. You can improve your teaching substantially by getting good and regular feedback about where your students are, what they are thinking, and how they are interpreting the information provided in your course.

> *Redish's fourth teaching commandment:* Find out as much as you can about what your students are thinking.

Four plausible and frequently used approaches to getting feedback from your students and evaluating their learning are:

1. Observe student behavior in class and in office hours.
2. Measure student satisfaction with a survey or questionnaire.
3. Measure student learning using a closed-ended question (multiple-choice or short-answer), designed using the results of physics education research on commonly found errors to specify attractive distractors.
4. Measure student learning using open-ended (long-answer or essay) exam questions—problems or open-expression questions in which students explain and discuss their answers.

The first approach is an essential part of understanding our students and seeing how they are responding to our instruction. Here, we have to be particularly careful not to go into wishful-thinking mode and assume that a question that a student asks in office hours simply needs to be answered in a direct and straightforward manner. This may sometimes be the case, but more often, I have found that asking a few well-placed questions such as "Well, what do *you* think is going on?" or "Why do you ask?" or even "Could you explain why you're stuck?" often produces a lot of information—and indicates that my "gut-response" answer would have been useless.

> *Redish's fifth teaching commandment:* When students ask you a question or for help, don't answer right away. Ask them questions first, in order to see whether your assumptions about their question are correct.

The second approach, an attitude survey or questionnaire, is the simplest and most commonly used, but although student satisfaction is important in motivating student work, and presumably therefore in producing student success, the link between satisfaction and learning is highly indirect. Indeed, students whose primary goal is a good grade may find higher satisfaction in a course that produces a good grade without improved learning, since improved learning often requires time and painful effort.

The third method, multiple-choice questions with research-based distractors or short-answer questions with research-motivated contexts, is easy to deliver, but, as explained in chapter 5, requires a substantial effort to develop effectively. I discuss four kinds of such questions below: multiple-choice with research-based distractors, multiple-choice multiple-response, representation translation, and ranking tasks. Other formats for short-answer questions have been developed and can be quite effective [Peterson 1989].

The fourth approach, long-answer or open-ended questions, is easy to deliver, but the grading and analysis of long-answer questions and problems can be time consuming. Student answers must be read in detail and evaluated by the understanding displayed. Grading here can be quite subtle. There is a tension between grading that is too draconian and only gives credit for detailed and careful reasoning, and grading that is too casual and gives points for remembered equations or principles that the student has no idea how to use. The former tends to produce grades that are too low and the latter to send the message that understanding is not important. Later in this chapter, I discuss four kinds of long-answer questions: open-ended or context-rich reasoning problems, estimation questions, qualitative questions, and essay questions.

TESTING

Often the standard testing we carry out to assess our students is limited, especially in large classes. In order to facilitate grading, we might choose to resort to machine-graded closed-answer questions—multiple-choice end-of-chapter problems that rely primarily on recognizing the correct equation and performing some straightforward algebraic manipulations to get a number. Decisions to test in this way have a powerful impact on our instruction. Not only do we severely limit the kind of feedback we receive about how our students are doing, but we send them a powerful message about what we believe is important for them to learn.

College students are as busy as their instructors, having many courses to consider, not just physics. In addition, many of them are young adults seeking partners and engaging in extramural sports or clubs, and some are working long hours to pay for their tuition. Some of these extramural activities will have more impact on their lives and careers than their physics class, hard as this might be to believe! As a result, even if I provide my students with good advice—such as that working something out in multiple ways is important for their learning, or that once they have finished a problem they should think about it to see that it makes sense—they are likely to ignore it if they can't see how doing it will pay off directly in points and grades. Most students regularly practice "time triage," doing only those activities that seem absolutely necessary for getting through the class successfully.[3]

In order to get students to engage in the kind of intellectual activity we want them to engage in, we have to let them know by testing them for it directly. In particular, in choosing exam questions, we have to be aware of the mixed messages we may be sending. This can be quite difficult, especially in these days where a class's success is measured by "student happiness" as evaluated with end-of-class student anonymous comments, as opposed to being measured by some assessment of student learning. Many students are satisfied with getting

[3]And of course "successfully" means different things to different students. For a few it means showing their parents that they were not cut out to be doctors or engineers or whatever goals their parents may have imposed on them.

through a class with a decent grade without having to do too much work—even if they learn little or nothing—and on end-of-semester surveys they may reward teachers who help them achieve this goal and punish those who make them work too hard.[4]

Designing exams

The kinds of questions we choose for an exam can make a big impact on what students choose to do in our classes. If, for example, we construct our exams from multiple-choice questions that can be answered by recognition (that is, there are no "tricky" or tempting alternatives), our students are likely to be satisfied with reading their text and lecture notes "lightly" a number of times. If we construct our exams from homework problems they have been assigned, our students are likely to seek out correct solutions from friends or solution books and to memorize them without making the effort to understand or make sense of them. If all our exam problems can be solved by coming up with the "right equation" and turning the crank, students will memorize lists of equations and pattern match. If we allow our students to create a card of equations and bring it into the exam and then only test the application of equations, students are likely to forego understanding altogether and only practice equation manipulation. These kinds of "studying" have minimal impact on real learning and are often highly ephemeral.

If we really want students to study physics deeply and effectively in a way that will produce long-term learning, the activities we provide are not enough; the learning they foster has to be validated via testing in their examinations.

> *Redish's sixth teaching commandment:* If you want your students to learn something, you have to test them on it. This is particularly true for items in the "hidden curriculum" (cf. chapter 3).

Exams as formative feedback

Exams and quizzes are not only means of carrying out summative assessments of what students have learned. They can also provide formative feedback to students as to what they have learned and what they need to work on. Unfortunately, it has been my experience that most students have the expectation that exams only serve as summative assessments. If they do poorly, they respond, "Well I sure messed that up. I need to do better on the next topic." Of course, since physics is highly cumulative, both in content and in skill building, having a significant "hole" in one's knowledge can lead into a downward spiral.

There are ways of designing the presentation and delivery of exams and quizzes to help encourage students to pay attention to the mistakes they make on exams. I've developed a pattern of exam delivery that seems to work reasonably well to encourage at least some students to debug their thinking from the feedback they get on exams.

- The exam is given in class during the last class of the week.
- The exam is graded immediately (over the weekend) and returned in the first class of the next week.

[4]Be careful! The interpretation that one's bad responses from students is because one is making them work too hard is a classic "wishful thinking."

- I return the exams to the students at the beginning of the class and go over the exam in class, explicitly showing the partial credit grading patterns, if any.
- I encourage the students to look for grading errors or possible interpretations of the problem's wording that would improve their grade. I tell them to submit a written request for a grade change with a detailed explanation of why they think they should have received more points. (Just saying "please look at problem 4" doesn't count. They have to explain why they think they were right and the grader was wrong.)
- If they are dissatisfied with their grade on the exam, they may take an out-of-class makeup test on the same material (but not the same test). However, the grade on the makeup does not replace the first test: they get the average of the two grades.
- I tell them that my experience is that a student who got a low grade who simply goes back and studies again in the same way (or does not study at all) is as likely to go down on the makeup and lose points as a result. On the other hand, students who specifically study their first exam in order to understand the mistakes they made and why have a high probability of raising their grade substantially.
- The class averages on each question are reported and they are told that the final exam will include at least one question based on the questions the class performed the most poorly on.

Since this procedure takes up two lecture periods for every exam and requires me to write two exams for every exam, I tend to reduce the number of hour exams I give in a semester: two instead of three in a course with three 50-minute lectures per week; one and two shorter quizzes in a course with two 75-minute lectures per week. In a lecture class of 100 to 200, I usually get about 25% of the students taking the makeup. This has been effective in making the students more comfortable with my unfamiliar exam style. (I always include essay questions, estimation questions, and representation translation questions, along with a more traditional problem or two.)

The last point in the bulleted list sends the message that I expect them to know everything I am testing them on. If the class as a whole performs badly, I review the physics behind the problem carefully when I go over it in class and tell them that they all need to go back and look at that material again—and that they are certain to have a question concerning that topic on the final exam.

EIGHT TYPES OF EXAM AND HOMEWORK QUESTIONS

A wide variety of structures are available for presenting physics problems to our students. Different types of structures tend to activate different kinds of associations and resources. A critical element in developing a course that gets students thinking more deeply about physics is choosing from a broad palette of question types. In the rest of this chapter, I discuss eight kinds of problems, briefly describe their value, and give an example or two of each. In addition to the kinds of problems discussed here, see also the discussion of Peer Instruction and JiTT in chapter 7 and of Cooperative Problem Solving in chapter 8 where other problem structures are used.

The Physics Suite has a large collection of problems, being the union of the set developed over the years for *Fundamentals of Physics* [HRW6 2001], the set of *Workshop Physics* problems, and the set I have developed over the years for the Activity-Based Physics project. These problems are distributed in the text, *Understanding Physics*, in the problem volume, and some are available on the Resource CD in the back of this book. Additional sources of problems and approaches to testing may be found in [Arons 1994], [Tobias 1997], and the books listed in the file "Other Resources" on the Resource CD associated with this volume.

Multiple-choice and short-answer questions

Multiple-choice and short-answer questions are tempting to use because they are easy to grade. The results can be highly suggestive, but multiple-choice tests can be difficult to interpret. They tend to overestimate the student's learning since they can sometimes be answered correctly by means of incorrect reasoning[5] or by cued responses that fail to represent functional understanding. On the other hand, the use of common misconceptions or facets as distractors produces "attractive nuisances" that challenge the students' understanding. Students who get the correct answer despite these challenges are likely to have a reasonably good understanding. A well-constructed multiple-choice question based on research into common naïve conceptions can therefore give some indication of the robustness of a student's correct answer. Note, however, that standard multiple-choice questions developed by instructors who have not studied the relevant research often have distractors that are too trivial to provide a real test of students' understanding. Instructors unaware of research results in physics education sometimes find it difficult to imagine the kinds of errors that students will commonly make.

An example of a good multiple-choice question with research-based distractors is shown in Figure 4.1. The distractors have to correspond to students' naïve conceptions, not to what faculty think. Since most physics instructors know very well that imbedding an object in a fluid produces a buoyant force, they may find the answers to the problem in Figure 4.1 peculiar. But

A book is at rest on a table top. Which of the following force(s) is (are) acting on the book?

1. A downward force due to gravity.
2. The upward force by the table.
3. A net downward force due to air pressure.
4. A net upward force due to air pressure.

(A) 1 only
(B) 1 and 2
(C) 1, 2, and 3
(D) 1, 2, and 4
(E) None of these. Since the book is at rest, there are no forces acting on it.

Figure 4.1 A multiple-choice question from the FCI [Hestenes 1992].

[5] See, for example, [Sandin 1985].

it is well documented that many high school students and some college students think that air pressure is responsible for gravity, pushing down on an object with the weight of the air above it, so item (C) is an attractive distractor for those students.[6] A large collection of multiple-choice questions with research-based distractors is provided in the conceptual surveys of various topics in physics given on the Resource CD associated with this volume.

Short-answer questions can also explore students' functional understanding of physics concepts effectively. The key is not to make the question a simple recognition test but to require some reasoning, perhaps using ideas or principles that are not directly cued by the problem. In the sample shown in Figure 4.2, the problems naturally cue up buoyant forces, but they require a fairly sophisticated application of free-body diagrams. (The answer to 2.2, for example, is obviously "=" since both net forces are equal to 0 by Newton's first law. Students who do not comfortably distinguish individual and net forces have trouble with this.)

Multiple-choice multiple-response questions

A substantial amount of thinking and reasoning can be required from a student in a multiple-choice multiple-response test. A sample is shown in Figure 4.3. To solve this problem, a

For each of the following partial sentences, indicate whether they are correctly completed by the symbol corresponding to the phrase *greater than* (>), *less than* (<), or *the same as* (=).

(2.1) A chunk of iron is sitting on a table. It is then moved from the table into a bucket of water sitting on the table. The iron now rests on the bottom of the bucket. The force the bucket exerts on the block when the block is sitting on the bottom of the bucket is _______ the force that the table exerted on the block when the block was sitting on the table.

(2.2) A chunk of iron is sitting on a table. It is then moved from the table into a bucket of water sitting on the table. The iron now rests on the bottom of the bucket. The total force on the block when it is sitting on the bottom of the bucket is _______ it was on the table.

(2.3) A chunk of iron is sitting on a table. It is then covered by a bell jar which has a nozzle connected to a vacuum pump. The air is extracted from the bell jar. The force the table exerts on the block when the block is sitting in a vacuum is _______ the force that the table exerted on the block when the block was sitting in the air.

(2.4) A chunk of iron is sitting on a scale. The iron and the scale are then both immersed in a large vat of water. After being immersed in the water, the scale reading will be _______ the scale reading when they were simply sitting in the air. (Assume the scale would read zero if nothing were sitting on it, even when it is under water.)

Figure 4.2 Sample short-answer question combining the use of different physics principles.

[6] Then, Sagredo asks, where does the weight of the air come from? A good question, but one not often asked by these students.

Four different mice (labeled A, B, C, and D) ran the triangular maze shown below. They started in the lower left corner and followed the paths of the arrows. The times they took are shown below each figure.

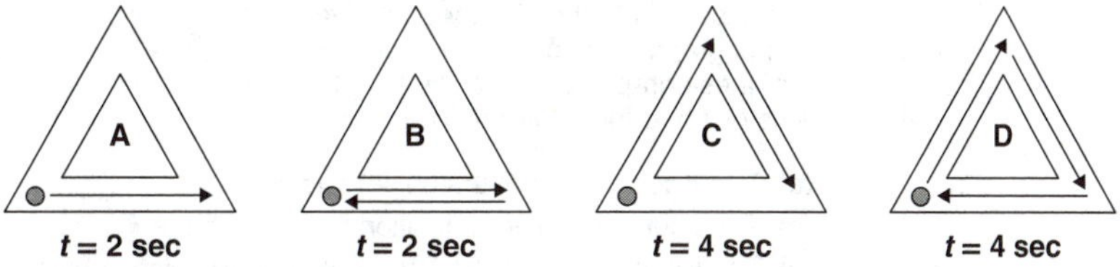

For each item below, on your answer sheet write the letters of all of the mice that fit the description.

(a) This mouse had the greatest average speed.
(b) This mouse had the greatest total displacement.
(c) This mouse had an average velocity that points in this direction (⇒).
(d) This mouse had the greatest average velocity.

Figure 4.3 A sample multiple-choice multiple-response question.

student has to have good control of the concepts of vector displacement and average vector velocity, and to be able to clearly distinguish velocity and speed.

This type of question requires a student to evaluate each statement and make a decision about it. It is a particularly useful type of question in cases where students tend to have mixed context-dependent models of a physical situation.

Representation-translation questions

As discussed in chapter 2, learning to handle the variety of representations we use can be quite a challenge for introductory students, but it can be one of the most valuable general skills they develop from studying physics. The presentation of a single situation in different ways facilitates understanding and sense-making of different facets of the situation. Furthermore, one of the primary difficulties novice students have with problem solving is their failure to be able to visualize the dynamic situation and map a physics description onto that visualization.

Thornton and Sokoloff [Thornton 1998] pioneered the use of problems in which students are required to match a statement in words describing a physical situation with graphs or pictures that describe those situations. You will find many such problems in their concept surveys contained in the Action Research Kit on the Resource CD: the FMCE, the ECCE, the VET, and the MMCE.

I have had success getting students to think about the meaning of physics variables with problems in which they are shown a situation and a number of graphs. The graphs have their abscissa labeled as time, but the ordinates are unmarked. The students then have to match a list of physical variables to the graphs. An example is shown in Figure 4.4.

I had an interesting experience with a student concerning the representation translation shown in Figure 4.5. After the exam in which this problem was given, the student (an

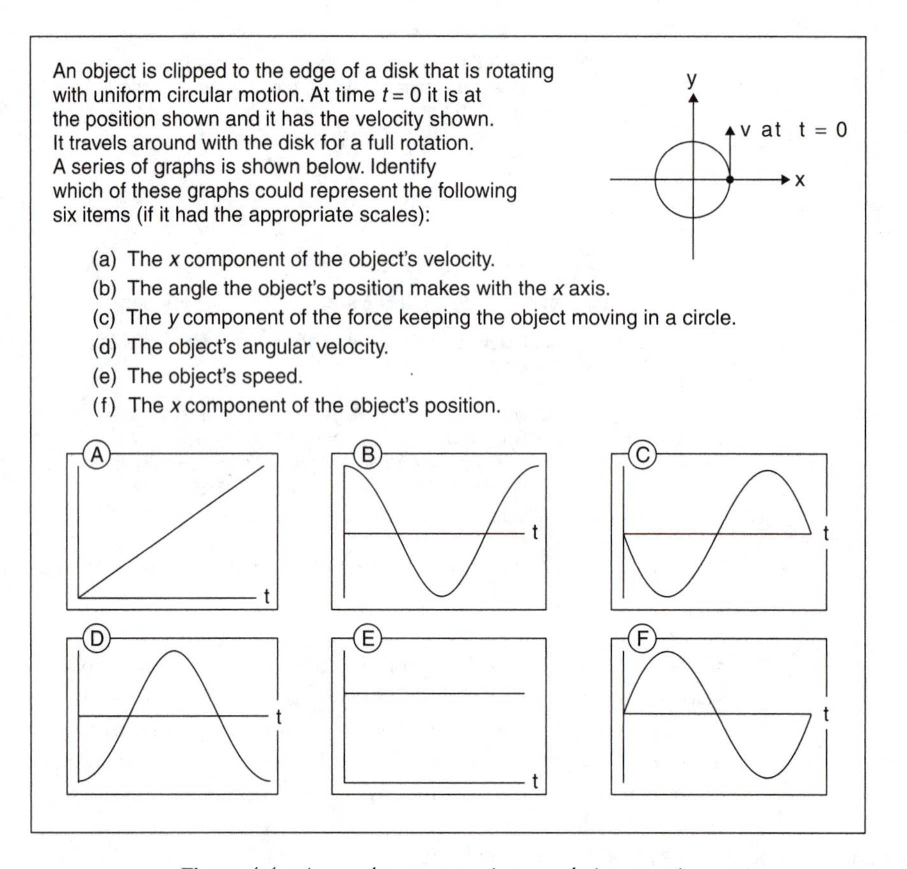

Figure 4.4 A sample representation-translation question.

engineer) came in to complain. "I don't understand how to do these problems," he grumped. I asked what answer he chose for part (a). He reported that he had selected option (e). "And why did you choose that one?" I asked. "Well," he said, "it's a wave and waves are supposed to be wiggly. That was the wiggliest."

"Okay," I responded. "Now tell me what's happening."

"What do you mean?" he asked.

I said, "Describe the string, tell me where the dot is. Then tell me what happens to the string and the dot as the pulse moves down the string."

"Hmmm," he said. "Well, the pulse moves to the right, when it gets to the dot, the dot moves up and then down . . . Oh, [expletive deleted]!" Once he had worked through visualizing what happened, he was almost trivially able to solve the rest of the problem.

Although I have not carried out explicit research on the topic, in my experience, the students in both algebra- and calculus-based physics struggle with these problems and the struggle is extremely productive.

Consider the motion of a pulse on a long taut string. We will choose our coordinate system so that when the string is at rest, the string lies along the x axis of the coordinate system. We will take the positive direction of the x axis to be to the right on this page and the positive direction of the y axis to be up. Ignore gravity. A pulse is started on the string moving to the right. At a time t_0 a photograph of the string would look like Figure A below. A point on the string to the right of the pulse is marked by a spot of paint.

For each of the items below, identify which figure below would look most like the graph of the indicated quantity. (Take the positive axis as up.) If none of the figures look like you expect the graph to look, write *N*.

(a) The graph of the y displacement of the spot of paint as a function of time.
(b) The graph of the x velocity of the spot of paint as a function of time.
(c) The graph of the y velocity of the spot of paint as a function of time.
(d) The graph of the y component of the force on the piece of string marked by the paint as a function of time.

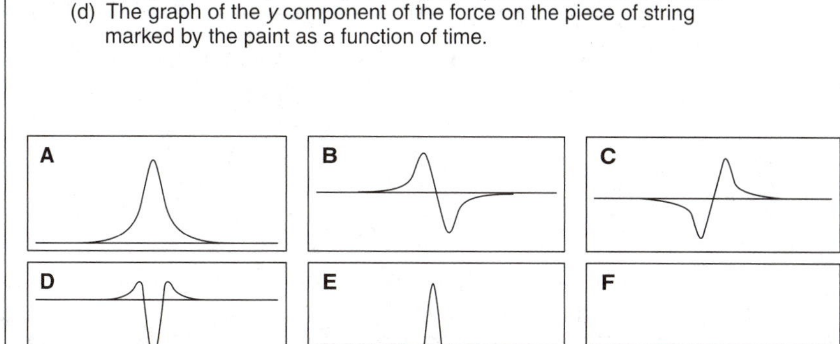

Figure 4.5 A sample representation-translation problem.

Ranking tasks

Another class of easy-to-grade but effective problems are *ranking tasks*—problems in which the student must order a series. These have been used by a number of researchers and curriculum developers effectively.

Ranking tasks are effective because they easily trigger reasoning primitives[7] such as "more of a cause produces (proportionately) more of an effect" [diSessa 1993]. An example of a ranking task from the UWPEG is shown in Figure 4.6.

David Maloney has been a long-time user of ranking tasks in his research. Recently, he and his collaborators, Tom O'Kuma and Curt Heiggelke, published a collection of these problems [O'Kuma 1999]. A sample is given in Figure 4.7. Many of the Reading Exercises in *Understanding Physics* are ranking tasks.

[7] Primitives are discussed in detail in chapter 2.

In the picture shown below are 5 blocks. The blocks have equal volumes but different masses. The blocks are placed in an aquarium tank filled with water, and blocks 2 and 5 come to rest as shown in the lower figure. Sketch on the figure where you would expect the blocks 1, 3, and 4 to come to rest. (The differences in mass between successive blocks is significant—not just a tiny amount.)

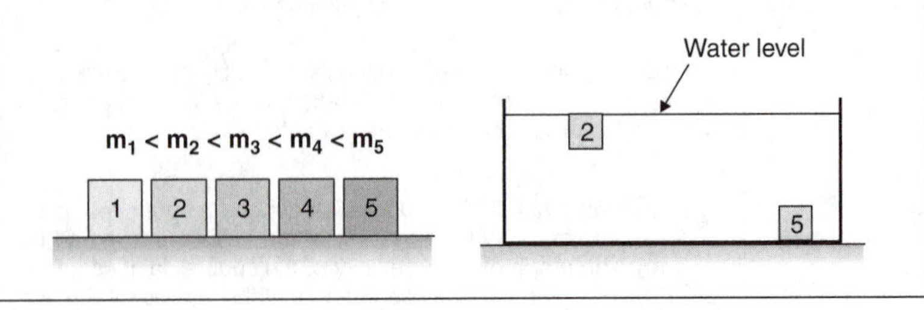

Figure 4.6 A ranking task [Loverude 1999].

Given below are seven arrangements of two electric charges. In each figure, a point labeled *P* is also identified. All of the charges are the same size, *Q*, but they can be either positive or negative. The charges and point *P* all lie on a straight line. The distances between adjacement items, either between two charges or between a charge and the point *P*, are all *x* cm. There is no charge at point *P*, nor are there any other charges in this region.

Rank these arrangements from greatest to least on the basis of the strength of the electric field at point *P*. That is, put first the arrangement that produces the strongest field at point *P*, and put last the arrangement that produces the weakest field at point *P*.

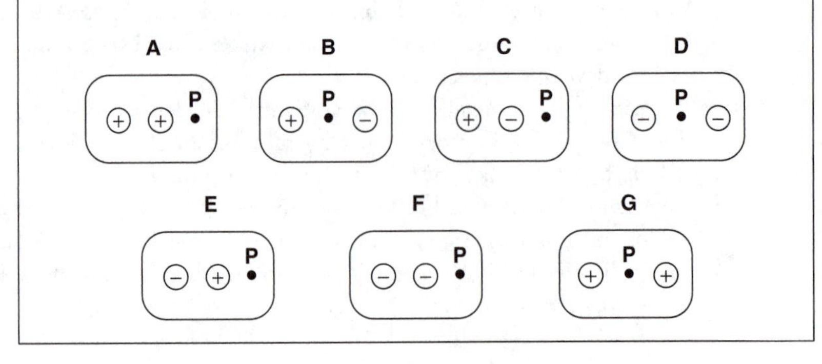

Figure 4.7 A ranking task from [O'Kuma 1999].

Context-based reasoning problems

The problems I find most valuable, both on homework and on exams, are those I call *context-based reasoning problems.*[8] In these problems, students are given a reasonably realistic situation and have to use physics principles—often in ways or circumstances in that they have not been previously seen—to come to a conclusion. The crucial fact is that the answer to the

I have set up my two stereo speakers on my back patio as shown in the top view diagram in the figure at the right. I am worried that at certain positions I will lose frequencies as a result of interference. The coordinate grid on the edge of the picture has its large tick marks separated by 1 meter. For ease of calculation, make the following assumptions:

- Assume that the relevant objects lie on integer or half-integer grid points of the coordinate system.
- Take the speed of sound to be 343 m/s.
- Ignore the reflection of sound from the house, trees, etc.
- The speakers are in phase.

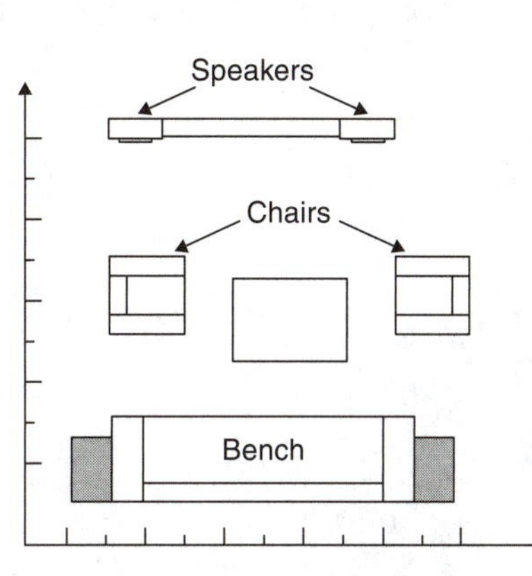

(a) What will happen if I am sitting in the middle of the bench?

(b) If I am sitting in the lawn chair on the left, what will be the lowest frequency I will lose to destructive interference? (If you do not have a calculator, leave the result as an expression with numbers that could be simply evaluated and estimate the result to one significant figure.)

(c) Can I restore the frequency lost in part (a) by switching the leads to one of the speakers, thereby reversing the phase of that source?

(d) With the leads reversed, what will happen to the sound for a person sitting at the center of the bench?

Figure 4.8 A context-based reasoning problem.

[8] These are similar in spirit to *Context-Rich Problems* used by the Minnesota group. See the discussion of Cooperative Problem Solving in chapter 8.

problem should be of some reasonable real-world interest. Unfortunately, too many end-of-chapter physics problems are poorly motivated. They seem simply an exercise in carrying out some obscure physics calculations for no obvious purpose. A problem such as "How much work is done by a weight-lifter in raising a 200-kg barbell a distance of 2 meters" is of this type. Why is the "work" something I should care about? What makes this calculation relevant? A problem such as "Estimate the number of calories a marathon runner burns in running 26 miles. Does he need to stock up on carbohydrates before beginning?" is better motivated.

An example that I like is shown in Figure 4.8. This is particularly nice since it relies on the fundamental idea of interference—figure out the path difference and see how many wavelengths fit in. The geometry does not permit the use of the small-angle approximation that leads to the standard interference formulas. Students have to calculate distances using the Pythagorean Theorem. It's not too realistic because with sound, reflections from nearby walls and objects are of primary importance. But this calculation is one part of understanding what is going on.

Everyday examples in newspapers, advertising, television, and movies in which the physics is wrong make nice problems of this type. An example is given in Figure 4.9.

In the movie *Jurassic Park,* there is a scene in which some members of the visiting group are trapped in the kitchen with dinosaurs on the other side of the door. The paleontologist is pressing his shoulder near the center of the door, trying to keep out the dinosaurs. The botanist throws her back against the door at the edge next to the hinge. She cannot reach a gun on the floor because she is trying to help hold the door closed. Would they improve or worsen their situation if he moved to the door's outer edge and she went for the gun? Estimate the change in the torque they would exert on the door if they changed their positions as suggested.

Figure 4.9 A context-based reasoning problem.

> Estimate the number of blades of grass a typical suburban house's lawn has in the summer.

Figure 4.10 A typical Fermi question.

Estimation problems

Estimation problems were made famous by Enrico Fermi, who was a master at them. His classic question was "How many barbers are there in Chicago?" Questions of this type can have considerable value because students

- get to practice and apply proportional reasoning
- learn to work with large numbers
- learn to think about significant figures (I always deduct points for too many significant figures.)
- learn to quantify their real-world experience

An example of a typical Fermi question is given in Figure 4.10. I use this question to explain to my class what it is that I want them to do when they solve problems of this type.[9] To get a number, students must picture a lawn, estimate its area, and estimate the number of blades of grass in a given area of lawn and then scale up. I always require that they start from a number they can plausibly know. Thus, if they said "let's assume that a square meter of lawn has a million blades of grass," I would give no credit for that estimate. If, on the other hand, they said,

> Consider a square centimeter of grass. There are about 10 blades of grass counting along each side. So there will be 100 in a square centimeter and 100 × 100 times that in a square meter. So assume one million blades of grass per square meter.

that answer would receive full credit. I always grade (or have graded) my estimation problems so that points are given for each part of the reasoning required. I explain carefully to my students that they are going to be evaluated on how they come up with the answer, not just on the answer.

An instructor needs to be both persistent and patient to include estimation problems. At first, students may not believe that you are serious about asking them to do problems of this type. For this reason, I give an estimation problem on every homework and exam, and I identify some homework estimations as having come from previous exams so that students

[9] This question is appropriate for Maryland, where the lawns in the summer still have grass. It might be too easy for a student at the University of New Mexico in Albuquerque.

will know I'm serious. Second, students may not believe that this is the sort of thing they are supposed to be learning in physics class and resent doing them at first. One student complained on his post-class anonymous survey that "exam points were given for being a good guesser." In my experience, this difficulty passes as they gain skill and confidence. Near the end of one of my algebra-based physics classes, one of my students told me with great glee, "Prof, you know those estimation problems? Well I'm taking a business class and we're doing, like, making a business plan and, you know, it's just like estimations? I was the only one in the class who knew how to do it!"

As the class moves on through the year, I begin to blend more physics into estimation problems so that they become design problems. These seem to play a big role in helping students understand the long-term value of physics for their professional future. An example of an "estimation/design" problem is given in Figure 4.11.

Qualitative questions

Qualitative questions can be quite effective in getting students to learn to think about concepts—and in helping instructors realize that students may not have learned what they thought they had from observing good performance on quantitative problem solving.[10] The University of Washington group has been particularly effective in using this kind of question in curriculum design.[11]

A sample qualitative problem that is an extension of a ranking task is given in Figure 4.12. I first encountered questions of this type when visiting the University of Washington on sabbatical in 1992. My inclination was to label each resistor, *R*, label each battery, *V*, write down all the equations using Kirchhoff's laws, and solve them for the relevant unknowns. This method certainly works! But one of the facilitators[12] asked me whether I could solve it without equations. "Why should I?" I responded. "Because," he said, "perhaps your students are not as facile with equations as you are." I proceeded to give it a try and was quite surprised at how difficult it was. I realized that I was using the physics equations as a scaffold for organizing my conceptual knowledge. For students who have not developed this knowledge structure and who might be hazy about what the concepts mean, my approach would be largely inaccessible. The approach of reasoning through the problems conceptually, on the other hand, helps reveal students' naïve conceptions (such as the idea that a battery is a source of constant current). As they work through these problems on homework and struggle with them on exams, they seem to build the concepts more firmly than they would if they could get away with a purely mathematical approach. Note that an essential part of questions of this type is the phrase "explain your reasoning." Many students don't know what reasoning or building a case means [Kuhn 1989]. Having them explain their reasoning, discussing what you mean by reasons, and giving them feedback on the sort of things we mean by reasoning can be an important part of what they learn in a physics class.

[10] See the discussion of Mazur's experience in chapter 1.

[11] See the discussion of Tutorials in Introductory Physics in chapter 8 and Physics by Inquiry in chapter 9.

[12] Richard Steinberg, with whom I later had the privilege of collaborating extensively.

You are assigned the task of deciding whether it is possible to purchase a desk-top-sized magnetic spectrometer in order to measure the ratio of C^{12} to C^{14} atoms in a sample in order to determine its age.

For this probem, let's concentrate on the magnet that will perform the separation of masses. Suppose that you have burned and vaporized the sample so that the carbon atoms are in a gas. You now pass this gas through an "ionizer" that on the average strips one electron from each atom. You then accelerate the ions by putting them through an electrostatic accelerator—two capacitor plates with small holes that permit the ions to enter and leave.

The two plates are charged so that they are at a voltage difference of ΔV Volts. The electric field produced by charges on the capacitor plates accelerate the ions to an energy of $q\Delta V$. These fast ions are then introduced into a nearly constant, vertical magnetic field. (See the figure below.) If we ignore gravity, the magnetic field will cause the charged particles to follow a circular path in a horizontal plane. The radius of the circle will depend on the atom's mass. (Assume the whole device will be placed inside a vacuum chamber.)

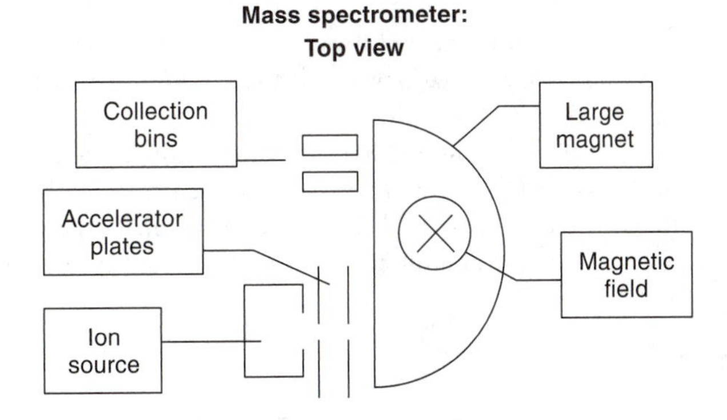

Answer three questions about how the device works.
(a) We would like not to use too high a voltage. If ΔV is 1000 Volts, how big a magnetic field would we require to have a plausible "table-top-sized" intrument? Is this a reasonable magnetic field to have with a table-top sized magnet?

(b) Do the C^{12} and C^{14} atoms hit the collection plate far enough apart? (If they are not separated by at least a few millimeters at the end of their path we will have trouble collecting the atoms in separate bins.)

(c) Can we get away with ignoring gravity? (*Hint:* Calculate the time it would take the atom to travel its semicircle and calculate how far it would fall in that time.)

Figure 4.11 An estimation/design problem.

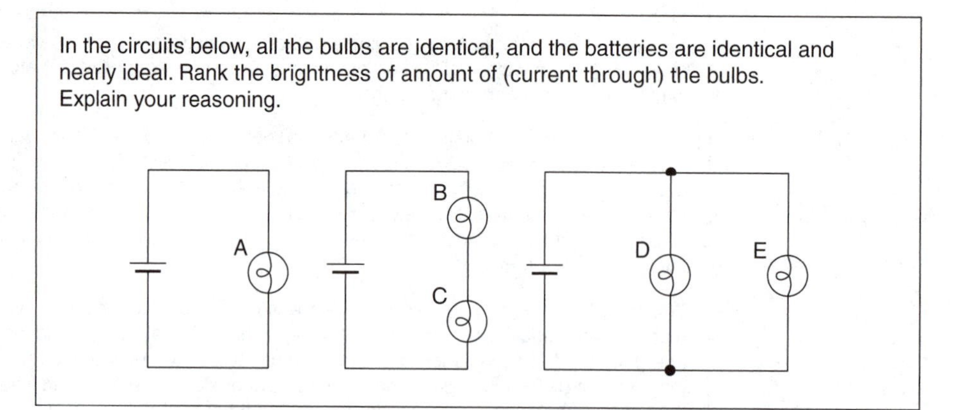

Figure 4.12 Sample qualitative question on direct current circuits [McDermott 1992].

Of course, in the end I want my students to develop a well-organized knowledge structure and to have equations mapped firmly into that structure. But it is important not to demand too much too early.

Qualitative questions that require identifying relevant physics principles and concepts, qualitative reasoning, and writing an explanation can be effective in helping students make the connection between their real-world personal experiences and the physics they are learning. An example of a question of this type is given in Figure 4.13.

Essay questions

Essay questions can be the most revealing of students' difficulties and naïve conceptions of any form. In the example given in Figure 4.14, the students are not asked to recall the law—it is given to them. But they are asked to discuss its validity. In this case, we had completed an ABP Tutorial (see chapter 8) exploring Newton's third law (N3) with two force probes. I had wanted them to refer to experimental evidence of some sort in justifying their belief in N3. Interestingly enough, a significant fraction of students did not expect N3 to hold in all

In public restrooms there are often paper towel dispensers that require you to pull downward on the towel to extract it. If your hands are wet and you are pulling with one hand, the towel often rips. When you pull with both hands, the towel can be extracted without tearing. Explain why.

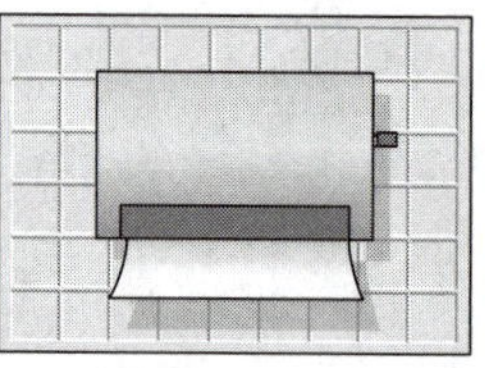

Figure 4.13 A sample qualitative reasoning problem that makes a connection to the student's personal experience.

Newton's 3rd law says that objects that touch each other exert forces on each other.

> *If object A exerts a force on object B, then object B exerts a force back on object A and the two forces are equal and opposite.*

Consider the following three situations concerning two *identical* cars and a much heavier truck.

(a) One car is parked and the other car crashes into it.
(b) One car is parked and the truck crashes into it.
(c) The truck is pushing the car because the car's engine cannot start. The two are touching and the truck is speeding up.

In which of these situations do you think Newton's 3rd law holds or does not hold? Explain your reasons for saying so. (Your answers are worth 5 points, your reasons 10.)

Figure 4.14 A sample essay question on Newton's 3rd law.

these cases. The results provided me guidance for follow-up discussions in lecture and later problems on exams. In the example given in Figure 4.15, I learned in a dramatic fashion that a large majority of my students had significant difficulties with the concept of "field" despite my careful attempts to be precise in lecture.

Note that giving essay questions in an exam situation has to be done with some care. There are three points to consider.

1. In some university environments (such as mine), introductory students often have all their classes as large lectures and they may have little or no experience with actually having to write during an examination.

In this semester we have studied two fields: electric and magnetic. Explain why we introduce the idea of field and compare and contrast the electric and magnetic fields. In your comparison, be certain to discuss at least one similarity and one difference.

Figure 4.15 A sample essay question on the field concept.

2. Remember that in exams students are under tremendous emotional pressure. If the exam is long so that they are also under time pressure, they may not have the opportunity to think about what they are writing.
3. The person who learns the most from essay questions on exams is the person grading them.

As a result of these points, I always try to keep exams short enough that the students are not overly rushed and I always grade my own essay questions, even if I have a large class and have TAs available for grading the other parts of the exam.

CHAPTER 5

Evaluating Our Instruction: Surveys

Mathematics may be compared to a mill of exquisite workmanship, which grinds you stuff of any degree of fineness; but, nevertheless, what you get out depends on what you put in; and as the grandest mill in the world will not extract wheat flour from peascod, so pages of formulae will not get a definite result out of loose data.

T. H. Huxley [Huxley 1869]

As I discussed in the last chapter, there are two ways to probe what is happening in one's class. One way is to assess how much each student has learned in order to decide the extent to which that student receives a public certification of his or her knowledge—a grade. A second way is to probe our class overall in order to determine whether the instruction we are delivering is meeting our goals. I refer to the first as *assessment*, and the second as *evaluation*. In the last chapter we discussed how to assess each student so as to see what he or she was learning. In this chapter, we discuss how to get a snapshot of how the class is doing overall for the purpose of evaluating our instruction.

As a result of the context dependence of the cognitive response (Principle 2 in chapter 2), in some contexts students may choose to use the model they are being taught, while in other contexts they may revert to using more naïve resources. When we look at a class broadly (especially a large class), we can tolerate larger fluctuations in individual responses than we can when we are assessing individual students. The students' individualized choices of what ideas to use and when to use them depend on uncontrollable and unknowable variables (their internal mental states). As a result, their answers may appear random on a small set of closed-end questions on each topic to be probed. But these same few questions may give a good average view of what is happening, despite not giving a complete picture of the knowledge of any single student.

RESEARCH-BASED SURVEYS

A cost-effective way to determine the approximate state of a class's knowledge is to use a carefully designed research-based survey. By a *survey* I mean a reasonably short (10- to 30-minute) machine-gradable test. It could consist of multiple-choice or short-answer questions, or it could contain statements students are asked to agree or disagree with. It can be delivered on paper with Scantron™ sheets or on computers.[1] It can be delivered to large numbers of students and the results manipulated on computers using spreadsheets or more sophisticated statistical analysis tools.

By *research-based*, I mean that the survey has been developed from qualitative research on student difficulties with particular topics and has been refined, tested, and validated by detailed observations with many students. Broadly, to achieve good surveys (surveys that are both valid and reliable—see the discussion below) requires the following steps.

- Conduct qualitative research to identify the student models underlying their responses.
- Develop a theoretical framework to model the student responses for that particular topic.
- Develop multiple-choice items to elicit the range of expected possible answers.
- Use the results—including the student selection of wrong answers—to facilitate the design of new instructions as well as new diagnostic and evaluation tools.
- Use the results to guide construction of new qualitative research to further improve the survey.

This process places development of evaluational tools firmly in the research-redevelopment cycle of curriculum construction and reform discussed in more detail in chapter 6 (Figure 6.1).

Surveys may focus on a variety of aspects of what students are expected to learn in both the explicit and hidden curriculum. *Content surveys* probe student knowledge of the conceptual bases of particular content areas of physics. *Attitude surveys* probe student thinking about the process and character of learning physics. Over the past two decades, physics education researchers have developed dozens of surveys that probe topics from mechanics (the Force Concept Inventory) to the atomic model of matter (the Small Particle Model Assessment Test). Seventeen such surveys are included on the Resource CD accompanying this volume. They are listed in the Appendix at the end of this volume.

Why use a research-based survey?

Sagredo scoffs at my emphasis on creating a survey through research. "I've given machine-gradable multiple-choice final exams in my large classes for years. Don't my grades count as course evaluations?" Certainly they do, Sagredo. But there are dangers in interpreting exam results as course evaluations.

The questions we choose often are constrained by a number of factors that may be unrelated to student learning. The first danger is that there is pressure from students (and sometimes from administrations) to have an "appropriate" grade distribution. Students are

[1] Studies to look for differences between paper-delivered and computer-delivered surveys have so far had ambiguous results.

comfortable with class averages near 80%, with 90% being the boundary for an A and with few grades falling below 60%. Although this may make students and administrations happy, it presses us to produce the requisite number of As no matter what our students have learned.[2]

A second danger arises because we are interested in what our students have really learned, not in what they think you want them to say. By the time they get to university, many students have become quite adept at "test-taking skills." These are even taught in some schools in order to help students (and school administrators) receive higher evaluations. I'm not criticizing students for taking this approach. I made use of them myself when I took standardized tests. Students taking an exam have the goal to obtain the highest possible score given what they know. Instructors want the score to accurately reflect what their students know. If we are not aware in detail of our students' starting states—what resources and facets are easily activated—we might be hard pressed to come up with reasonable wrong answers for a multiple-choice test or with tempting and misleading cues for short-answer questions. Without these *attractive distractors,* students can focus on eliminating obviously wrong answers and can use their test-taking skills to get a correct result even if they only have a very weak understanding of the subject.

Neither of these dangers is trivial. The other side of the first danger is that instruction is never purely objective. It is oriented, in principle, to achieving goals set by the instructor, though sometimes those goals are tacit, inconsistent, or inappropriate to the particular student population involved. An instructor's exams should reflect his or her own particular learning goals for his or her students. What is appropriate for us to demand our students learn is a continuing negotiation among instructors, their students, and outside pressures such as administrators, parents, and faculty in the departments our courses serve.

If an instructor is unaware of common student confusions or of how students tend to respond to particular questions, the result on questions she creates for an exam may not reflect what she thinks it does. Furthermore, without a carefully developed question based on research and a clear understanding of common naïve responses, the students' wrong answers may provide little useful information beyond "my students don't know the answer."

In trying to interpret the responses of students on closed exam questions, we may encounter one or more of the following problems.

1. If a multiple-choice test does not have appropriate distractors, you may not learn what the students really think.
2. The fact that students give the right answer does not mean they understand why the answer is right.
3. Since student responses are context dependent, what they say on a single question only tells part of the story.
4. Problems in the ordering or detailed presentation of the distractors may cause problems in interpreting the results.

[2] In my classes, I try to set exams that have averages between 60% and 65%. A grade over 75% is considered an A. At this level of difficulty, even the good students get some feedback about where they need to improve. See my model of examination delivery discussed in chapter 4.

5. It's easy to overinterpret the implications of a single relatively narrow test—especially if it only has one style of question.

Concept surveys that are carefully constructed with these points in mind can provide a useful tool as part of our evaluation of our instruction.

Surveys and the goals of a class

While giving a lecture on physics education research to colleagues in a neighboring physics department, I once showed some of the questions from the Force Concept Inventory (discussed in detail below and given on the Resource CD). One faculty member objected quite vigorously. "These are trick questions," he said. "What do you mean by a 'trick question'?" I asked. He answered, "You really have to have a deep understanding of the underlying physics to answer them correctly." After a substantial pause, allowing both him and the rest of the audience to consider what had just been said, I responded, "Exactly. I want all the questions I ask to be trick questions."

This raises a deep question. What is it we want our students to learn from our instruction? My colleague clearly had much lower expectations for his students than I did—in one sense. He was satisfied with recognition of an answer but didn't care if his students could not distinguish between the correct (physics) answer and an attractive (but incorrect) common-sense alternative. On the other hand, he probably demands much more in the way of sophisticated mathematical manipulations on examinations than I do and is satisfied if his students can match a complex problem-solving pattern to one they have memorized. I do not care if my students can pattern match complex mathematical manipulations. I want them to be able both to formulate physics problems out of real-world situations and to interpret their answers sensibly. If we could attain both goals in a one-year course, I would be delighted, but at present, I don't know how to do it given the time and resource constraints of a large class.

This shift in goals can produce some difficulties. Sagredo and I both teach algebra-based physics on occasion. When he looks at my exams, he complains that they are too easy and that I'm "dumbing-down" the course. Interestingly enough, many students have reported to me that the scuttlebutt among the students is "take Sagredo if you want an easy A" and that my course is the one to take "if you want to work hard and really understand it." Whenever Sagredo agrees to give one of my questions to his students on an exam, he is surprised at how poorly they do. My students would also do poorly on some of his questions.

In the end, when the chalk meets the blackboard, each individual instructor defines his or her own goals. Nonetheless, there is clearly a need for a community to form to discuss both the appropriate goals for physics instruction and how to evaluate the extent to which those goals are reached. That is why I favor the use of research-based surveys as one element in our evaluations of our instructional success. They are explicit in what they are evaluating, they are based on careful study of student difficulties, and they are carefully tested for validity and reliability.

Delivering a survey in your class

Whenever possible, I give pre-post surveys (i.e., at the beginning and end of the class). In some classes that have an associated lab, the first and last weeks of class do not have labs, and so I tell students to come in then to take surveys. In classes that do not have a lab or a blank week, I am willing to take time in the first and last classes to give surveys. To encourage students to take them (and to make them up if they miss the class), I give everyone who completes each survey 5 grade points (out of a total of about 1000 points for the class as a whole). If everyone does them, the surveys have no impact on the grading pattern.

To analyze a survey, it is important to compare only *matched data sets*. That is, only students who take both the pre- and the post-tests should be included. This is because there may be biases in the populations who take the two tests. For example, students who drop the course in the middle would take the pre-test and not the post-test. If the group of students dropping the class were biased toward the lower scoring students on the pre-test, this would bias the pre-post comparison toward high gains. At least if a matched set is used, one is looking at the true gains for a particular set of students.

The danger discussed in the previous section—that students often give us what they think we want instead of what they think—has three important implications for how surveys should be delivered, especially if the survey is to meet the purpose of evaluating our instruction rather than certifying the students. The three implications are:

- We have to be careful to "teach the physics" but not "teach to the test."
- Survey solutions should not be distributed or "gone over" with the students.
- Surveys should be required (given credit) but not graded.

The first implication, not teaching to the test, is a delicate one. We want the test to probe students' knowledge appropriately, and we want our instruction to help them gain the knowledge that will be probed. Why then is "teaching to the test" usually considered such a pejorative? I think that it is because in this case we are implicitly using a fairly sophisticated model of student learning: students should learn how to think, not to parrot back answers they don't understand. In our cognitive model (described in chapter 2), this idea can be expressed more explicitly by saying that students should develop a strong mental model of the physics with many strong associations that will permit them to identify and use appropriate solution techniques to solve a wide variety of problems presented in diverse contexts. Research strongly demonstrates that when students learn an answer to a problem narrowly, through pattern matching, small changes in the problem's statement can lead to their being unable to recognize the pattern.[3] So if during instruction we give students the specific question that will appear on the test, framed exactly as it will be framed there, I call it *teaching to the test.*

This leads to the second implication: Survey solutions should not be posted or given out to students. Research-based survey items can take a long time to develop. Students have to be interviewed to see how they are reading and interpreting the items and their answers. Surveys have to be delivered repeatedly to study distributions and reliability at a cost to class time. A carefully developed survey, whatever limitations it may have, is an extremely valuable resource for the community of physics teachers, but it is fragile. If they are graded and the answers are posted or discussed in class, they spread—to fraternity/sorority solution

[3] I have seen students who solved problems by pattern matching fail to recognize a problem they knew if the picture specifying the problem was reversed (mirror image).

collections and to student websites—and become a property of the student community rather than of the instructional community. They are transformed from a moderately effective evaluation tool for the teacher to a "test-taking skills" tool for the student.

This leads directly to the third implication: Surveys should not be a part of the student's grade. This is a somewhat controversial point. Sagredo suggests that students will not take a test seriously unless it is graded. This may be true in some populations. At Maryland, it has been my experience that 95% of my students hand in surveys that show they have been thought through and answered honestly.[4] This might differ in other populations. If a test is graded, at least some students will make a serious effort to find out the correct answers and perhaps spread them around. Since I very much don't want my students to do this, I treat my exams (which I consider pedagogical tools to help facilitate student learning) and my surveys (which I consider evaluational tools to help me understand my instruction) differently.

There is an additional reason for leaving surveys ungraded. Students often use their test-taking skills to try to produce an answer that they think the teacher will like, even if they don't really think that is the answer. A graded exam definitely tends to cue such responses. I am more interested in finding out how students respond when such motivation is removed in order to see whether instruction has had a broader impact on student thinking.

For a survey to be useful, it should be both valid and reliable. I discuss these conditions next. In the remainder of the chapter I discuss two kinds of surveys that have been developed and that are currently widely available: content surveys and attitude surveys.

UNDERSTANDING WHAT A SURVEY MEASURES: VALIDITY AND RELIABILITY

In order to be a useful tool in evaluating instruction, a survey should be *valid*; that is, it should measure what it claims to measure. A survey should also be *reliable;* that is, it should give reproducible results. When we're talking about measurements of how people think about physics instead of about measurements of physical properties, we have to consider carefully what we mean by these terms.

Validity

Understanding the validity of a survey item, either in a content or attitude survey, is not as trivial as it may appear on the surface. What's in question is not just the issue of whether the physics is right, but whether the question adequately probes the relevant mental structures. To see what this means, consider the following example. The most common student confusion about velocity graphs is whether they should "look like the motion" or like the rates of change of the motion. If you ask students in introductory university physics "which graph corresponds to an object moving away from the origin at a uniform rate" (as in the problem shown in Figure 5.4) and provide them with the correct (constant) graph but <u>not</u> with the choice of the attractive distractor (the linearly increasing graph), you will get a high score but an invalid question. This one is especially subtle. Jeff Saul found that if both of these graphs

[4] Evidence to the contrary might be: all answers the same, answers in a recurring pattern, survey completed in one-fourth the average time, and so on.

were included but the correct answer given first, ~80% of the students selected the right answer. But if the attractive distractor (a rising straight line) was given first, the success rate fell by about a factor of 2 [Saul 1996].

In order to achieve validity, we need to understand quite a bit not only about the physics but also about how students think about the physics. Since human beings show great flexibility in their responses to situations, we might expect an intractably large range of responses. Fortunately, in most cases studied, if the topic is reasonably narrowly defined, a fairly small number of distinct responses (two to ten) accounts for a large fraction of the answers and reasonings that are found in a given population. Understanding this range of plausible variation is absolutely essential in creating valid survey items. As a result, the first step in creating a survey is to study the literature on student thinking and do extensive qualitative research to learn how students think about the subject.

Even if we understand the range of student thinking on a topic, for an item to be valid students must respond to it in the expected way. This is one of the most frustrating steps in creating a survey. A good way to probe this is to observe a large number of students "thinking aloud" while doing the survey. Culture shifts and vocabulary differences between the (usually middle-aged) test designers and (usually young adult) subjects can produce dramatic surprises. In our validation interviews for the Maryland Physics Expectations (MPEX) survey discussed later in this chapter, we wanted to know if students understood our motivation for doing derivations in lecture. To our dismay we learned that a significant fraction of our calculus-based introductory physics students were unfamiliar with the word "derivation" and thought it meant "derivative."[5] To get a consistent valid response we had to rephrase our items.

Reliability

Reliability also has to be considered carefully. In probing human behavior, this term replaces the more standard physics term *repeatability*. We usually say that repeatibility means that if some other scientists repeated our experiment, they would get the same result, within expected statistical variations. What we really mean is that if we prepare a new experiment in the same way as the first experimenter did, using equivalent (but not the same) materials, we would get the same result. We don't expect to be able to measure the deformability of a piece of metal many times using the same piece of metal. We are comfortable with the idea that "all muons are identical," so that repeating a measurement of muon decay rates doesn't mean reassembling the previously measured muons out of their component parts.

But when it comes to people, we are accustomed to the idea that people are individuals and are not equivalent. If we try to repeat a survey with a given student a few days later, we are unlikely to get the identical result. First, the "state of the student" has changed somewhat as a result of taking the first survey. Second, the context dependence of the cognitive response reminds us that "the context" includes the entire state of the student's mind—something over which we have little control. Experiences between the two tests and local situations (Did a bad exam in another class make her disgruntled about science in general? Did an argument with

[5]This is perhaps a result of the unfortunate strong shift away from "proof" in high school math classes that took place in the 1990s.

his girl friend last night shift his concerns and associations?) may affect student responses. And sometimes, students do, in fact, learn something from thinking about and doing a survey.

Fortunately, these kinds of fluctuations in individual responses tend to average out over a large enough class. Measures can become repeatable (reliable) when considered as a measure of a population rather than as a measure of an individual. However, we must keep in mind that according to the individuality principle (Principle 4, chapter 2), we can expect a population to contain substantial spreads on a variety of measures. Any survey is measuring a small slice of these variables. There is likely to be a significant spread in results, and the spread is an important part of the data.[6]

In the educational world, reliability testing is sometimes interpreted to mean that students should respond similarly to the same question formulated in different ways. For example, one might present an "agree–disagree" item in both positive and negative senses. For the MPEX, discussed below, we have the following pair of items to decide whether a student feels that she needs to use her ordinary-world experiences in her physics class.

> *To understand physics, I sometimes think about my personal experiences and relate them to the topic being analyzed.*
>
> *Physics is related to the real world and it sometimes helps to think about the connection, but it is rarely essential for what I have to do in this course.*

Although these items are designed to be the reverse of each other, they are not identical. The difference between "sometimes" and "rarely essential" can lead a student to split the difference and agree with both items, especially if that student is on the fence or is in transition. Even when items are closer than these, students can hold contradictory views. In this case, the "lack of reliability" in the responses to the matched questions lies not in the test but in the student. Care must be taken <u>not</u> to eliminate questions that show this kind of "unreliability," lest one bias the survey toward only seeing topics on which most students have formed coherent mental models.

CONTENT SURVEYS

In the remainder of this section I discuss three of the most commonly used surveys in mechanics in detail: the Force Concept Inventory (FCI), the Force and Motion Conceptual Evaluation (FMCE), and the Mechanics Baseline Test (MBT). The additional content surveys that are included on the CD provided with this book are listed and described briefly in the Appendix.

The FCI

One of the most carefully researched and most extensively used concept surveys in our current collection is the Force Concept Inventory (FCI) developed by David Hestenes and his collaborators at Arizona State University [Hestenes 1992a]. This is a 30-item multiple-choice

[6] A useful metaphor for me is a spectral line. For many physical circumstances, the width and shape of the line is important data, not just its centroid.

survey meant to probe student conceptual learning in Newtonian dynamics. It focuses on issues of force (though there are a few kinematics questions), and it is easily deliverable. Students typically take 15 to 30 minutes to complete it.

Building a probe of student conceptual understanding requires both understanding the fundamental issues underlying the physics to be probed, as viewed by the professional scientist, and understanding the common naïve conceptions and confusions the students spontaneously bring to the subject as a result of their experience. In creating their mechanics concept test, Hestenes and his collaborators first thought carefully about the conceptual structure of Newtonian mechanics. But understanding the professional's view is not enough. The test has to be designed to respond properly when considered from the *student's* point of view. The distractors (wrong answers) should distract! That is, there should be answers that correspond to what many naïve students would say if the question were open ended and no answers were given.

Hestenes and his collaborators relied on existing research and did extensive research of their own to determine spontaneous student responses [Halloun 1985a] [Halloun 1985b] [Hestenes 1992a]. They then compiled a list of common naïve conceptions and attempted to create questions that would reveal whether or not the students harbored these naïve conceptions. Their list of naïve conceptions is given in the FCI paper [Hestenes 1992a]. Many of them are directly related to the facets created applying primitive reasoning in the context of motion. (See chapter 2.)

Finally, in constructing the FCI, Hestenes and his collaborators chose to use semirealistic situations and everyday speech rather than technical physics speech in order to set the context to be the student's personal resources for how the world works rather than what one is supposed to say in a physics class. See, for example, the upper part of Figure 2.6, and Figure 5.1. Answer (C) in Figure 5.1 is an example of a research-based distractor. Few physics instructors who have not studied the research literature would think of choosing such an item;

Imagine a head-on collision between a large truck and a small compact car. During the collision:

(A) the truck exerts a greater amount of force on the car than the car exerts on the truck.
(B) the car exerts a greater amount of force on the truck than the truck exerts on the car.
(C) neither exerts a force on the other; the car gets smashed simply because it gets in the way of the truck.
(D) the truck exerts a force on the car but the car does not exert a force on the truck.
(E) the truck exerts the same amount of force on the car as the car exerts on the truck.

Figure 5.1 A question from the Force Concept Inventory [Hestenes 1992a].

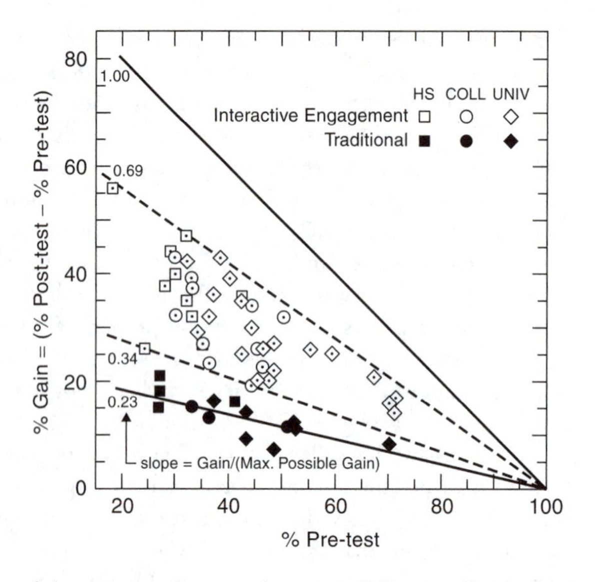

Figure 5.2 A plot of class average pre-test and post-test FCI scores for a collection of classes in high school, college, and university physics classes using a variety of instructional methods [Hake 1992].

it's not even "on their screen" as a possible wrong answer. But a significant number of naïve students, unaccustomed to seeking forces from objects as the cause of all changes in motion, actually select this answer.[7]

The FCI is perhaps the most widely used concept survey in the nation today. Its publication in 1992 stirred a great deal of interest in physics education reform among the community of college and university physics instructors. Looking at the test, most faculty declared it "trivial" and were shocked when their students performed poorly.[8] A typical score for a class of entering calculus-based physics students is 40% to 50% and a typical score for a class of entering algebra-based physics students is 30% to 45%. At the completion of one semester of mechanics, average scores tend to rise to about 60% for calculus-based students and 50% for algebra-based students. These are rather modest and disappointing gains.

Richard Hake of Indiana University put out a call for anyone who had given the FCI pre-post in a college class to send him their results, together with a description of their class. He collected results from over 60 classes [Hake 1992]. His results are displayed in Figure 5.2 and show an interesting uniformity. When the class's gain on the FCI (post-test average–pre-test average) is plotted against the class's pre-test score, classes of similar structure lie approximately along a straight line passing through the point (100,0). Traditional classes lie on the line closest to the horizontal axis and show limited improvement. The region between

[7]College students who have previously taken high school physics are less likely to choose (C) as an alternative here. They are more likely to select an "active agent" primitive or a facet built on a "more is more" primitive and to select (A).

[8]Compare the Mazur story in chapter 1. Mazur was influenced by the earlier version of the Hestenes test [Halloun 1985a], as was I.

the two dotted lines represents classes with more self-reported "active engagement." Hake claims that classes lying near the line falling most steeply reported that they were using active-engagement environments and a research-based text. This suggests that the negative slope of the line from a data point to the point (100,0) is a useful figure of merit:

$$g = (\text{class post-test average} - \text{class pre-test average})/(100 - \text{class pre-test average})$$

where the class averages are given in percents.

The interpretation of this is that two classes having the same figure of merit, g, have achieved the same *fraction of the possible gain*—a kind of educational efficiency. Hake's results suggest that this figure of merit is a way of comparing the instructional success of classes with differently prepared populations—say, a class at a highly selective university with entering scores of 75% and a class at an open enrollment college with entering scores of 30%. This conjecture has been widely accepted by the physics education research community.[9]

Hake's approach, though valuable as a first look, leaves some questions unanswered. Did people fairly and accurately represent the character of their own classes? Did a selection occur because people with good results submitted their data while people with poor results chose not to? To answer some of these questions, Jeff Saul and I undertook an investigation of 35 classes at seven different colleges and universities [Redish 1997] [Saul 1997]. Four different curricula were being used: traditional, two modest active engagement reforms (Tutorials and Group Problem Solving: ~one hour of reform class per week), and a deeply reformed high active-engagement curriculum (Workshop Physics) in early implementations.[10] We gave pre-post FCI in each class and observed the classes directly. The FCI results are summarized in Figure 5.3.

These results confirm Hake's observations and give support to the idea that g is one plausible measure of overall gain. Some additional interesting conjectures may be made after studying this figure.

1. In the traditional (low-interaction lecture-based) environment, what the lecturer does can have a big impact on the class's conceptual gains.

The peak corresponding to the traditional class is very broad. At Maryland, where we observed the largest number of classes in a reasonably uniform environment, the classes with the largest gains were taught by award-winning professors who tried to actively engage their classes during lecture. The lowest gains were taught by professors who had little interest in qualitative or conceptual learning and focused their attention on complex problem solving. (For the detailed "unsmoothed" version of the Maryland results, see Figure 8.3.)

2. In the moderate active-engagement classes (one modified small-class group-learning hour per week), much of the conceptual learning relevant to FCI gains was occurring in the modified class.

[9] Some detailed preliminary probes of this question have been reported at meetings of the American Association of Physics Teachers (AAPT), but no decisive publications have yet resulted.

[10] See chapters 8 and 9 for detailed descriptions of the methods.

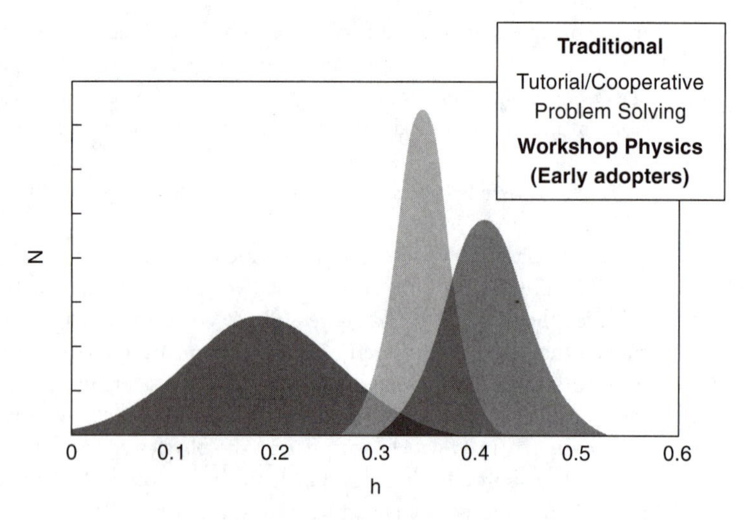

Figure 5.3 A plot of the fractional FCI gain achieved in three types of classes: traditional, moderate active engagement (tutorial/group problem solving), and strong active engagement (early adopters of workshop physics). Histograms are constructed for each group and fit with a Gaussian, which is then normalized [Saul 1997].

This is suggested by the narrowness of the peak and the fact that it lies above the results attained by even the best of the instructors in the traditional environments.

3. Full active-engagement classes can produce substantially better FCI gains, even in early implementations.

This is suggested by the results from the Workshop Physics classes studied. For a more detailed discussion of this issue (and for the results from mature Workshop Physics at the primary site), see chapter 9.

Although the FCI has been of great value in "raising the consciousness" of the community of physics teachers to issues of student learning, it has its limitations. Besides those limitations associated with all machine-gradable instruments, it is lacking in depth on kinematics issues that are to some extent a prerequisite to understanding the issues probed. A more comprehensive survey is provided by the Force and Motion Conceptual Evaluation.

The FMCE

The Force and Motion Conceptual Evaluation (FMCE) was developed by Ron Thornton and David Sokoloff [Thornton 1998]. In addition to the dynamical issues stressed by the FCI, this survey addresses student difficulties with kinematics, especially difficulties with representation translation between words and graphs. It is longer than the FCI, with 47 items in a multiple-choice multiple-response format that is somewhat more difficult for students to untangle than the (mostly) straightforward FCI multiple-choice items. As a result, students need more time to complete the FMCE—from 30 minutes to an hour.

An example of an FMCE item is given in Figure 5.4. Although this looks superficially trivial, students have a great deal of difficulty in choosing the correct graphs until they have

Questions 40–43 refer to a toy car which can move to the right or left along a horizontal line (the positive portion of the distance axis). The positive direction is to the right.

Choose the correct velocity-time graph (**A** - **G**) for each of the following questions. You may use a graph more than once or not at all. If you think that none is correct, answer choice **J**.

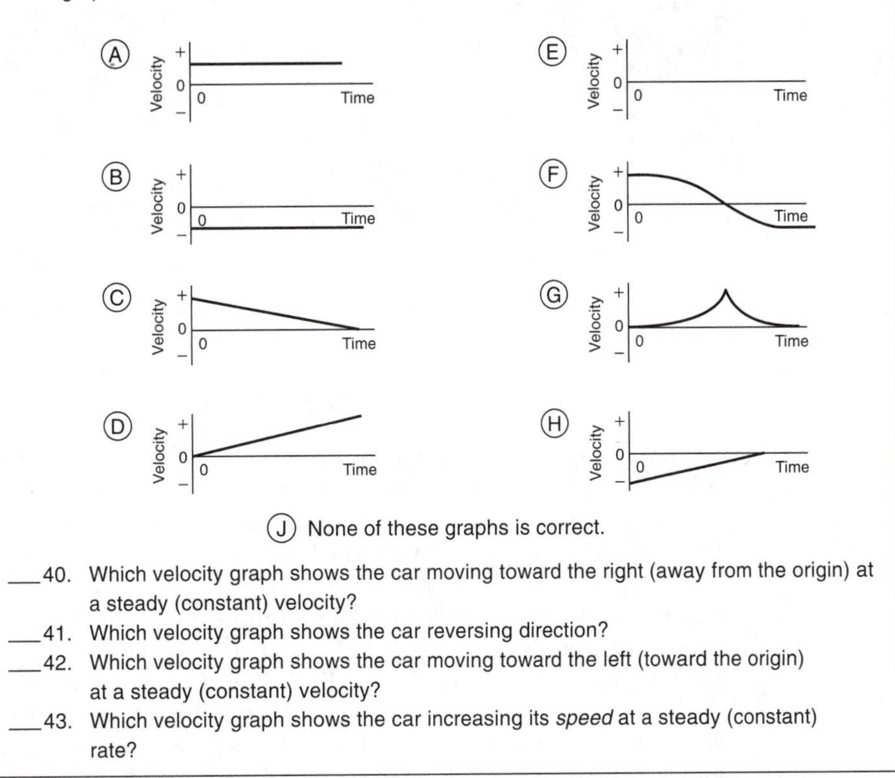

(J) None of these graphs is correct.

___40. Which velocity graph shows the car moving toward the right (away from the origin) at a steady (constant) velocity?

___41. Which velocity graph shows the car reversing direction?

___42. Which velocity graph shows the car moving toward the left (toward the origin) at a steady (constant) velocity?

___43. Which velocity graph shows the car increasing its *speed* at a steady (constant) rate?

Figure 5.4 An MCMR set of items from the FMCE [Thornton 1998].

clearly differentiated the concepts of velocity and acceleration and have developed good graph-mapping skills.[11] (In my own use of this survey, I exchange graphs (A) and (D) so as to better probe how many students are tempted to assign the "linearly rising" graph to a constant velocity.)

The FMCE is structured into clusters of questions associated with a particular situation, as shown in Figure 5.4. This tends to "lock" students into a particular mode of thinking for the cluster of problems and may not give a clear picture of the range of student confusion on a particular topic [Bao 1999].

[11] By "graph-mapping" skills, I mean the ability to map a physical situation onto a variety of different graphical representations.

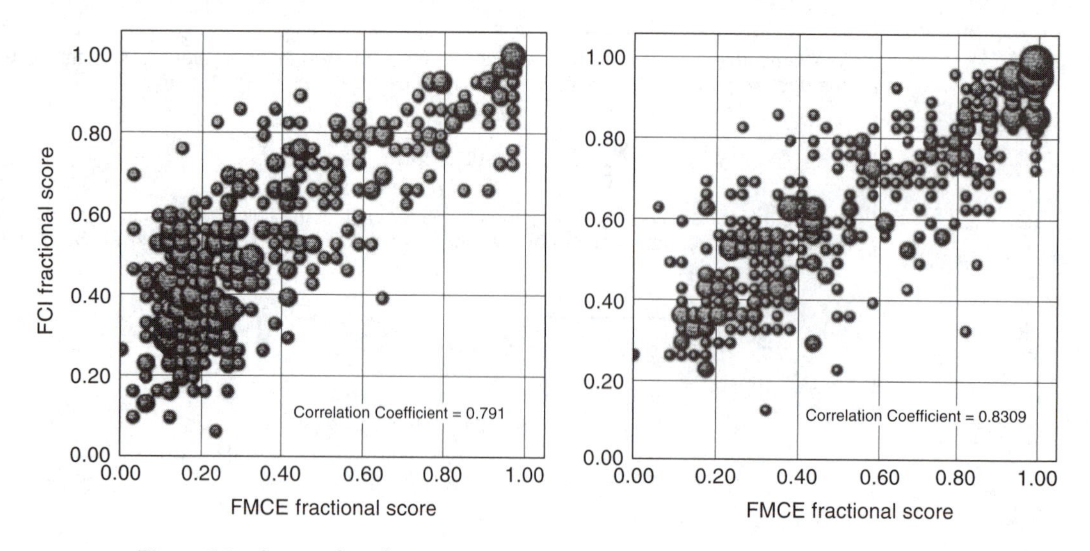

Figure 5.5 Scatter plot of FMCE versus FCI scores pre (left) and post (right). The size of the markers indicates the number of students with those scores [Thornton 2003].

Some of the items in the FMCE are included to "set up" the student's frame of mind or to check for students who are not taking the test seriously (e.g., item 33). All students are expected to get these correct since they cue widely held facets that lead to the correct result. See the description of the FMCE analysis in the file on the CD. (The FMCE analysis template included on the CD is already set up to handle this.)

Thornton and his collaborators have carried out extensive studies of the correlation between FMCE results and FCI results [Thornton 2003]. They find a very strong correlation ($R = 0.8$) between the results, but the FMCE appears to be more challenging to low-scoring students, with few students scoring below 25% in the FCI while FMCE scores go down to almost 0%. Scatter plots of pre- and post-FCI versus FMCE scores are shown in Figure 5.5. (The areas of the circles are proportional to the number of students at the point.)

The MBT

Both the FCI and the FMCE focus on components of basic conceptual understanding and representation translation. Our goals for physics classes at the college level usually include applying conceptual ideas to solve problems. Hestenes and his collaborators created the Mechanics Baseline Test (MBT) to try to probe students' skill at making these connections [Hestenes 1992b]. Scores tend to be lower than on the FCI. Although this survey is designed for the introductory physics class, David Hestenes told me that when he gave it to the physics graduate students in his first-year graduate classical mechanics class, it correlated well with their grades. An example of an MBT item is given in Figure 5.6. In this item, students have to recognize the relevance of energy conservation. Students who fail to do so tend to activate various facets or other associations.

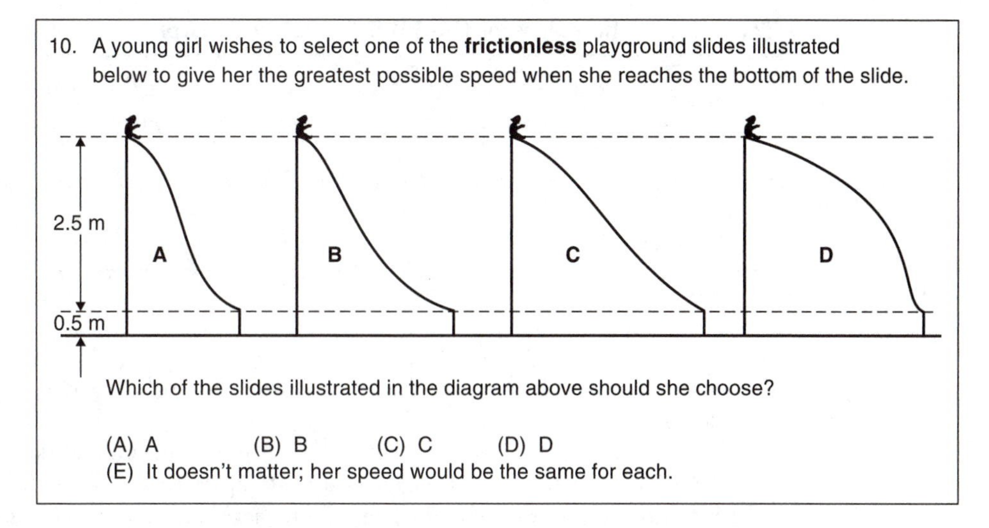

Figure 5.6 An item from the MBT [Hestenes 1992b].

ATTITUDE SURVEYS

If we want to understand whether our students are making any progress on our hidden curriculum of learning both process and scientific thinking, we need to find some way to probe the state of their attitudes.[12] One approach that has provided a useful first look is to use an attitude survey. Three attitude surveys are provided on the CD accompanying this book: the MPEX, the VASS, and the EBAPS.

In using an attitude survey, one needs to be aware of some limitations and caveats. First, attitudes, like most thinking process, are complex and context dependent. But they may fluctuate more widely than narrower content knowledge topics. The attitudes toward learning that students bring to our classroom may vary from day-to-day, depending on everything from whether they attended a party instead of studying the night before to whether a professor in another class has given a difficult and time-consuming homework assignment. Second, students' understanding of their own functional attitudes may be limited. Surveys of attitudes only measure what students think they think. To see how they really think, we have to observe them in action.

The MPEX

We created the Maryland Physics Expectations (MPEX) survey in the mid-1990s to provide a survey that could give some measure of what was happening to our students along the dimensions of relevance to the hidden curriculum [Redish 1998]. The focus of the survey was not on students' attitudes in general, such as their epistemologies or beliefs about the nature of science and scientific knowledge, but rather on their *expectations*. By expectations we mean that we want the students to ask themselves: "What do I expect to have to do in order to

[12] See chapter 3 for more discussion of the hidden curriculum.

TABLE 5.1 The Items of the MPEX Reality Cluster.

#10:	*Physical laws have little relation to what I experience in the real world.* (D)
#18:	*To understand physics, I sometimes think about my personal experiences and relate them to the topic being analyzed.* (A)
#22:	*Physics is related to the real world and it sometimes helps to think about the connection, but it is rarely essential for what I have to do in this course.* (D)
#25:	*Learning physics helps me understand situations in my everyday life.* (A)

succeed in this class?" I emphasize the narrowness of this goal: "this class," not "all my science classes" or "school in general."

The MPEX consists of 34 statements with which the students are asked to agree or disagree on a 5-point scale, from strongly agree to strongly disagree.[13] The MPEX items were validated through approximately 100 hours of interviews, listening to students talk about each item, how they interpreted it, and why they chose the answer they did. In addition, the parity of the favorable MPEX responses was validated by offering it to a series of expert physics instructors and asking what answers they would want their students to give on each item [Redish 1998]. The desired parity (agree or disagree) is labeled the *favorable* response, and the undesired parity is labeled *unfavorable.*

To illustrate the MPEX focus on expectations, consider the items given in Table 5.1. The favorable response (agree = A or disagree = D) is indicated at the end of the item. These items ask students to evaluate the link between physics and their everyday experience in two ways: from the class to their outside experience and from their outside experience to the class. Each direction is represented by two elements: one to which the favorable response is positive and one to which it is negative. The more general item "Physics has to do with what happens in the real world" was omitted, since almost all students agreed with it.

MPEX results

We analyze and display the results on the MPEX by creating an *agree/disagree (A/D) plot.* (See Figure 5.7.) In this plot, "agree" and "strongly agree" are merged (into "A"), and "disagree" and "strongly disagree" are merged (into "D"). The result is a three-point scale: agree, neutral, and disagree. This collapse of scale is based on the idea that, while it might be difficult to compare one student's "strongly agree" to another student's "agree," or to make much of a shift from "strongly agree" to "agree" in a single student, there is a robust difference between "agree" and "disagree" and a shift from one to the other is significant. The unfavorable responses are plotted on the abscissa and the favorable responses on the ordinate. Since the A + D + N ("neutral") responses must add to 100%, the point representing a class lies within the triangle bounded by the abscissa (unfavorable axis), the ordinate (favorable axis), and the line representing 0% neutral choices (F + U = 100). The expert responses are plotted as a cross in the favorable corner of the triangle.

[13] In education circles, such a ranking is referred to as a *Likert* (Lĭk-ert) scale.

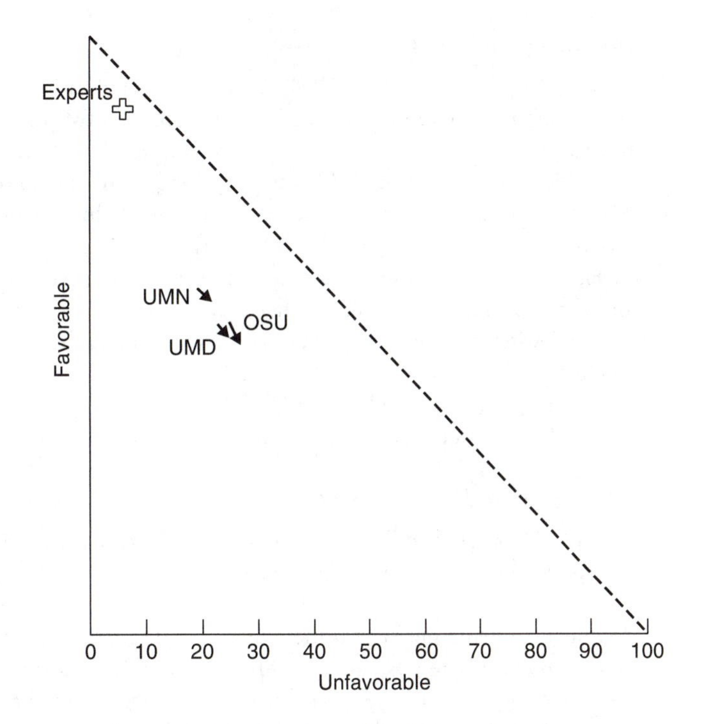

Figure 5.7 An agree/disagree (A/D) plot of overall MPEX results in large university settings at the beginning and end of a one-semester calculus-based physics class. All classes had traditional lectures with a one-hour active-engagement small-class session per week. Each represents data from ~500 students [Redish 1998].

Pre- and post-scores can be plotted for each cluster on the A-D plot. It is convenient to connect the pre- and post-results for each cluster with an arrow. A spreadsheet that allows you to paste in your MPEX results and generate A-D plots is included on the Resource CD associated with the volume.[14] Overall data from three large state universities are shown in Figure 5.7 (data from [Redish 1998]).

The MPEX has been delivered as a pre-post survey to thousands of students around the United States. The results have been remarkably consistent.

1. On the average, college and university students entering calculus-based physics classes choose favorable response on approximately 65% of the MPEX items.
2. At the end of one semester of instruction in large lecture classes, the number of favorable responses drops by approximately 1.5σ. This is true even in classes that contain active-engagement elements that produce significantly improved conceptual gains as measured, say, by the FCI.

[14] This spreadsheet was created by Jeff Saul and Michael Wittmann.

Analyzing the MPEX

Among the 34 items of the MPEX, 21 are associated into five clusters corresponding to the three Hammer variables described in chapter 3, plus two more. The five MPEX clusters are described in Table 5.2.

Note that some of the MPEX items belong to more than one cluster. This is because the MPEX variables are not interpreted as being linearly independent. The breakdown into clusters of the MPEX results at Maryland in the first semester of engineering physics is shown in Figure 5.8. These results are averaged over seven instructors and represent a total of 445 students (matched data). We see that there are significant losses in all of the clusters except concepts where we had a small gain. Note that not all of the MPEX items have been assigned to clusters.

Items 3, 6, 7, 24, and 31 (an "Effort cluster") are included to help an instructor understand what the students actually do in the class. They include items such as

> #7: *I read the text in detail and work through many of the examples given there.* (A)
>
> #31: *I use the mistakes I make on homework and on exam problems as clues to what I need to do to understand the material better.* (A)

Although students' answers to these items are interesting, I recommend that they not be included in overall pre-post analyses. There is a strong tendency for students to hope that they will do these things before a class begins, but they report that they didn't actually do them after the class is over. Inclusion of these items biases the overall results toward the negative.

MPEX items 1, 5, 9, 11, 28, 30, 32, 33, and 34 are not assigned to clusters. Interviews suggest that these items are indeed correlated with student sophistication, but they do not correlate nicely into clusters. Furthermore, since the MPEX was designed for a class in calculus-based physics for engineers, some of these items may not be considered as desirable goals for other classes.

TABLE 5.2 The MPEX Variables and the Assignment of Elements to Clusters.

	Favorable	Unfavorable	MPEX Items
Independence	Takes responsibility for constructing own understanding	Takes what is given by authorities (teacher, text) without evaluation	8, 13, 14, 17, 27
Coherence	Believes physics needs to be considered as a connected, consistent framework	Believes physics can be treated as unrelated facts or independent "pieces"	12, 15, 16, 21, 29
Concepts	Stresses understanding of the underlying ideas and concepts	Focuses on memorizing and using formulas without interpretation or "sense-making"	4, 14, 19, 23, 26, 27
Reality	Believes ideas learned in physics are relevant and useful in a wide variety of real contexts	Believes ideas learned in physics are unrelated to experiences outside the classroom	10, 18, 22, 25
Math link	Considers mathematics as a convenient way of representing physical phenomena	Views the physics and math independently with little relationship between them	2, 8, 15, 16, 17, 20

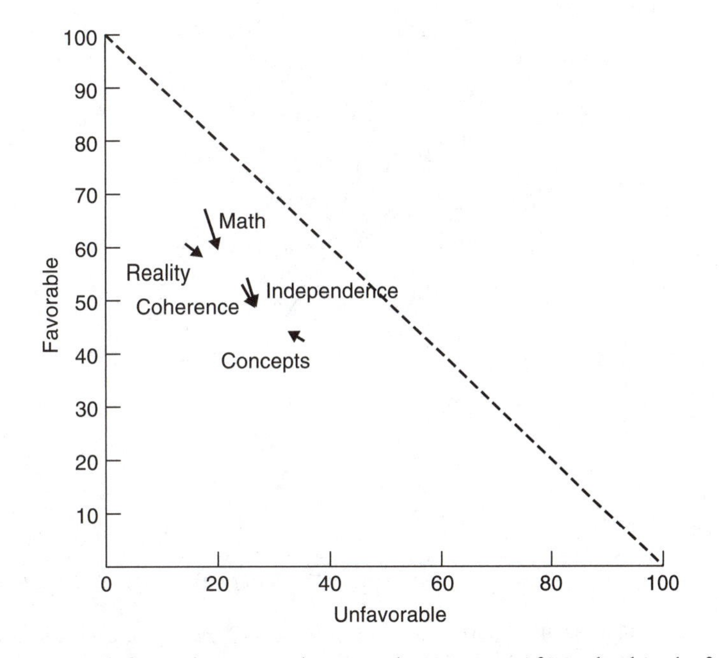

Figure 5.8 Pre-post shifts on the MPEX clusters at the University of Maryland in the first semester of engineering physics (data from [Redish 1998]).

Two items in particular tend to be controversial.

> #1: *All I need to do to understand most of the basic ideas in this course is just read the text, work most of the problems, and/or pay close attention in class.* (D)
>
> #34: *Learning physics requires that I substantially rethink, restructure, and reorganize the information that I am given in class and/or in the text.* (A)

Sagredo is unhappy about these. He says, "For #1, I would be happy if they did that. Why do you want them to disagree? For #34, some of my best students don't have to do this to do very well in my class. Why should they agree?" You are correct, Sagredo, and I suggest that you give these items, but not include them in your analysis or plots.[15] We include them because our interviews have revealed that the best and most sophisticated students in a class who are working deeply with the physics respond favorably as indicated. Certainly, for introductory courses this level of sophistication may be unnecessary. I like to retain these items, hoping that something I am doing is helping my students realize that developing a deep understanding of physics requires the responses as indicated.

Getting improvements on the MPEX

The fact that most courses probed with the MPEX show losses is discouraging but not unexpected. It is not surprising that students do not learn elements of the hidden curriculum

[15] This is easily achieved in the Saul-Wittmann MPEX analysis spreadsheet by replacing "1"s by "0"s in the appropriate cells.

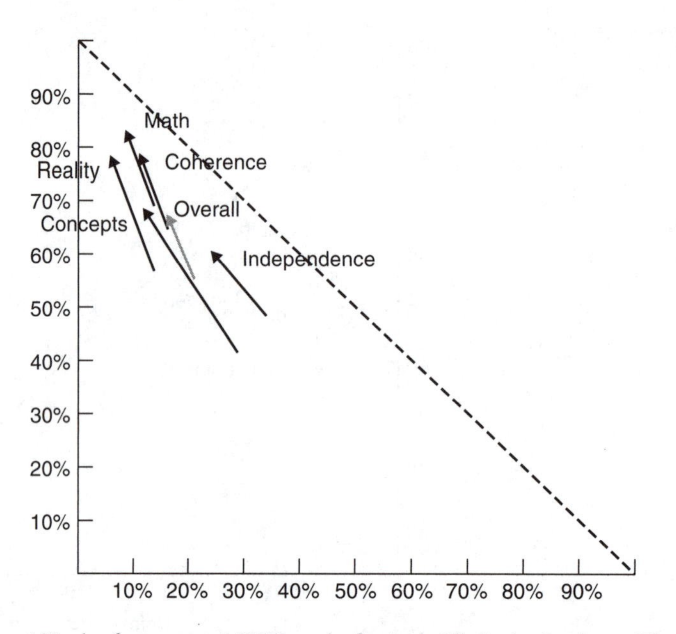

Figure 5.9 An A/D plot for pre-post MPEX results for Andy Elby's physics class at Thomas Jefferson High School in Virginia. For each cluster, the pre-result is at the base of the arrow, the post is at the tip of the arrow, and the name of the associated cluster is next to the arrowhead. The overall result is shown as a gray arrow [Elby 2001].

as long as it stays hidden. If we want students to improve along these dimensions, we have to be more explicit in providing structures to help them learn them.

Recently, MPEX results in classes designed to focus on explicit instruction in intuition building, coherence, and self-awareness of one's physics thinking have shown substantial improvements in all the MPEX categories [Elby 2001]. These results are shown in Figure 5.9.

I have been able to achieve MPEX gains in my large lecture classes by making specific efforts to keep issues of process explicit in both lectures and homework. Soon after developing the MPEX in 1995, I made strong efforts in my calculus-based physics classes to produce gains by giving estimation problems to encourage a reality link, by talking about process, and by stressing interpretation of equations in lecture. Results were disappointing. The responses on the reality link items still deteriorated, as did overall results. After much thought and effort, I introduced activities in lecture to help students become more engaged in these issues (see the discussion of Interactive Lecture Demonstrations in chapter 8), and I expanded my homework problems to include context-related problems every week. I reduced the number of equations I used and stressed derivations and the complex application of the few conceptually oriented equations that remained. The results (in my algebra-based class in 2000) were the first MPEX gains I had ever been able to realize.[16] Some of the results for four interesting items are shown in Table 5.3.

[16] This class also produced the largest FCI/FMCE gains I have ever managed to achieve.

TABLE 5.3 Pre- and Post-results on Four MPEX Items from a Calculus-Based Class Using UW Tutorials and Algebra-Based Class Using More Explicit Self-Analysis Techniques.

			Calculus-based (1995)			Algebra-based (2000)		
			F	U	N	F	U	N
#4	"Problem solving" in physics basically means matching problems with facts or equations and then substituting values to get a number.	Pre	60%	21%	19%	66%	30%	4%
		Post	77%	13%	10%	91%	9%	0%
#13	My grade in this course is primarily determined by how familiar I am with the material. Insight or creativity has little to do with it.	Pre	54%	24%	22%	57%	40%	3%
		Post	49%	23%	28%	79%	19%	2%
#14	Learning physics is a matter of acquiring knowledge that is specifically located in the laws, principles, and equations given in class and/or in the textbook.	Pre	39%	28%	33%	36%	53%	11%
		Post	37%	24%	39%	64%	34%	2%
(#19)	The most crucial thing in solving a physics problem is finding the right equation to use.	Pre	43%	32%	25%	45%	45%	10%
		Post	46%	26%	28%	72%	26%	2%

The MPEX serves as a sort of "canary in the mine" to detect classes that may be toxic to our hidden curriculum goals. The fact that most first-semester physics classes result in a deterioration of favorable results is telling. The fact that gains can be obtained by strong and carefully thought out efforts suggests that the use of the MPEX can be instructive, when judiciously applied.

The VASS

A second survey on student attitudes toward science was developed by Ibrahim Halloun and David Hestenes [Halloun 1996]. The Views about Science Survey (VASS) comes in four forms: one each for physics, chemistry, biology, and mathematics. The physics survey has 30 items. Each item offers two responses, and students respond to each item on an eight-point scale as shown in Figure 5.10. (Option 8 is rarely chosen.) In addition to items that probe what I have called expectations, the survey includes items that attempt to probe a student's epistemological stance toward science. A sample item is given in Figure 5.11.

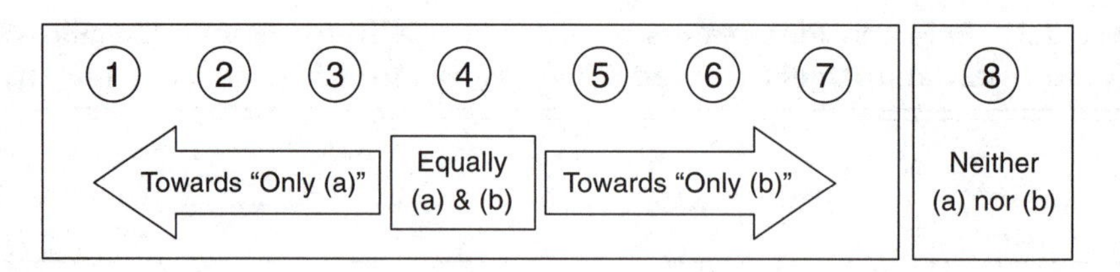

Figure 5.10 The eight-point scale for responding to items from the VASS [Halloun 1996].

The VASS is designed to probe student characteristics on six attitudinal dimensions—three scientific and three cognitive.

Scientific dimensions of the VASS

1. *Structure of scientific knowledge:* Science is a coherent body of knowledge about patterns in nature revealed by careful investigation rather than a loose collection of directly perceived facts (comparable to MPEX coherence cluster).
2. *Methodology of science:* The methods of science are systematic and generic rather than idiosyncratic and situation specific; mathematical modeling for problem solving involves more than selecting mathematical formulas for number crunching (extends MPEX math cluster).
3. *Approximate validity of scientific results:* Scientific knowledge is approximate, tentative, and refutable rather than exact, absolute, and final (not covered in the MPEX).

Cognitive dimensions of the VASS

4. *Learnability:* Science is learnable by anyone willing to make the effort, not just by a few talented people, and achievement depends more on personal effort than on the influence of teacher or textbook.
5. *Reflective thinking:* For a meaningful understanding of science one needs to concentrate on principles rather than just collect facts, look at things in a variety of ways, and analyze and refine one's own thinking.
6. *Personal relevance:* Science is relevant to everyone's life; it is not of exclusive concern to scientists (relates in part to MPEX reality cluster).

The favorable polarization of the VASS responses was determined by having it filled out by physics teachers and professors. Teachers' responses were strongly polarized on most items.

The laws of physics are:
(a) inherent in the nature of things and independent of how humans think.
(b) invented by physicists to organize their knowledge about the natural world.

Figure 5.11 A sample item from the VASS [Halloun 1996].

TABLE 5.4 Categories for Classifying VASS Responses [Halloun 1996].

Profile Type	Number of Items out of 30
Expert	19 or more items with expert views
High Transitional	15–18 items with expert views
Low Transitional	11–14 items with expert views and an equal or small number with folk views
Folk	11–14 items with expert views but a larger number of items with folk views or 10 items or less with expert views

Answers agreeing with the teachers are called *expert views,* while the polar opposites are referred to as *folk views*. Halloun and Hestenes classify students into four categories depending how well their responses agree with those of experts (see Table 5.4).

Halloun and Hestenes delivered the VASS to over 1500 high school physics students in 39 schools (30 of which used traditional rather than active engagement methods) at the beginning of class. They found them to be classified about 10% expert, about 25% high transitional, about 35% low transitional, and about 30% folk. Surveys of beginning college physics students gave similar results. For the high school students, there was a significant correlation between the students' profiles on the VASS and their gains on the FCI, as shown in Figure 5.12.

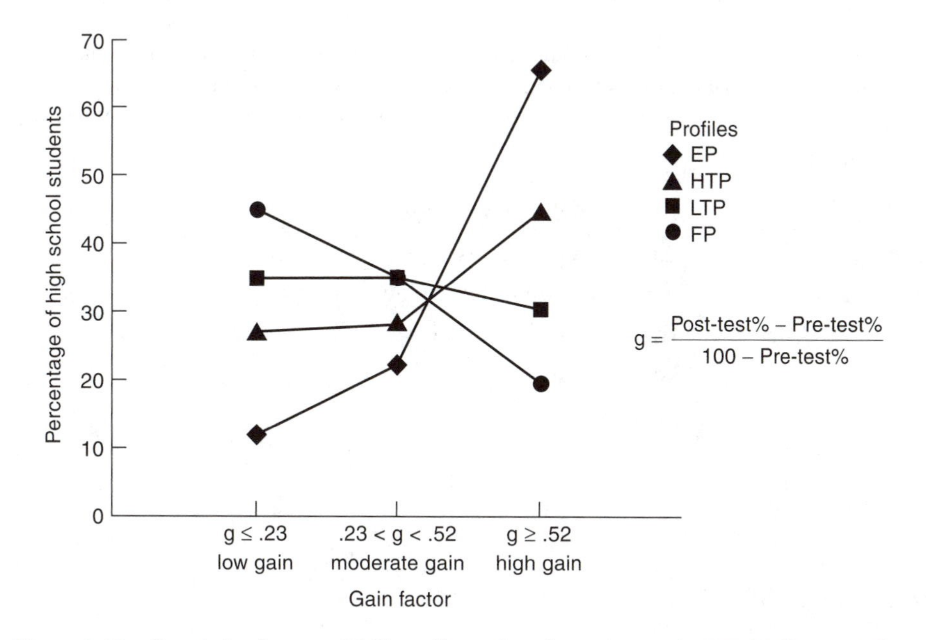

Figure 5.12 Correlation between VASS profiles and student gains on the FCI [Halloun 1996].

The EBAPS

Both the MPEX and the VASS suffer from the problem of probing what students think they think rather than how they function. In addition, they have the problem that for many items, the "answer the teacher wants" is reasonably clear, and students might choose those answers even if that's not what they believe. In the Epistemological Beliefs Assessment for Physics Science (EBAPS), Elby, Fredericksen, and White attempt to overcome these problems by presenting a mix of formats, including Likert-scale items, multiple-choice items, and "debate" items. Many EBAPS items attempt to provide context-based questions that ask students what they would do rather than what they think. The debate items are particularly interesting. Here's one.

> #26:
>
> **Justin:** When I'm learning science concepts for a test, I like to put things in my own words, so that they make sense to me.
>
> **Dave:** But putting things in your own words doesn't help you learn. The textbook was written by people who know science really well. You should learn things the way the textbook presents them.
>
> (a) I agree almost entirely with Justin.
>
> (b) Although I agree more with Justin, I think Dave makes some good points.
>
> (c) I agree (or disagree) equally with Justin and Dave.
>
> (d) Although I agree more with Dave, I think Justin makes some good points.
>
> (e) I agree almost entirely with Dave.

The EBAPS contains 17 agree-disagree items on a five-point scale, six multiple-choice items, and seven debate items for a total of 30. The Resource CD includes the EBAPS, a description of the motivations behind the EBAPS, and an Excel template for analyzing the results along five axes:

> Axis 1 = Structure of knowledge
>
> Axis 2 = Nature of learning
>
> Axis 3 = Real-life applicability
>
> Axis 4 = Evolving knowledge
>
> Axis 5 = Source of ability to learn

CHAPTER 6

Instructional Implications: Some Effective Teaching Methods

In theory, there is no difference
between theory and practice.
But in practice, there is.
Jan L. A. van de Snepscheut
quoted in [Fripp 2000]

The information in previous chapters describing what we know about the general character of learning and about the general skills we are trying to help students develop has profound implications for building an effective instructional environment. Sagredo once asked me, "OK. You've told me all this stuff about how people learn and shown me lots of references about specific student difficulties with particular bits of physics content. Now tell me the best way to teach my physics class next term."

Sorry, Sagredo. I wish it were that straightforward. First, as I pointed out in chapter 2, no single approach works for all students. Both individual differences and the particular populations in a class need to be taken into account. Second, despite the great progress in understanding physics learning that has been made in the past two decades, we're still a long way from being able to be prescriptive about teaching. All we can give are some guidelines and a framework for thinking about what might work for you. Third, the decisions teachers (or a department) make about instruction depend very strongly on the particular goals they would like to achieve with a particular course. Traditionally, these goals have been dominated by surface features rather than by deep structure—by selecting specific content matched, perhaps, to the long-term needs of the population being addressed rather than by thinking about student learning and understanding. The education research described above allows us to expand our community's discussion about what different students might learn from taking a particular physics course. This discussion has only just begun, and it is really only in the context of such a discussion that specific optimized curricula can be developed. Our goal is to transform good teaching from an art that only a few can carry out to a science that many can learn, but we've not gotten that far yet.

The traditional approach to physics at the college level involves lectures with little student interaction, end-of-chapter problem solving, and cookbook labs. Although students who are self-motivated independent learners with strong mathematical and experimental skills thrive in this environment (as they do in almost any educational environment), this category represents only a small fraction of our students. Indeed, the group seems to be shrinking, since young people today rarely have the "hands-on" mechanical experience common to physicists "of a certain age" and their teachers. The self-motivated independent learners of today are much more likely to have created their own computer games than to have built a crystal radio, rebuilt the engine of their parents' Ford, or been inspired by Euclid's *Elements*.

At present, we not only know a lot about where and why students run into difficulties, but the community of physics educators has developed many learning environments that have proven effective for achieving specific goals. With the Physics Suite, we pull together and integrate a number of these environments. In this chapter, I give brief overviews of innovative curricular materials that have been developed in conjunction with careful research, including both Suite elements and other materials that work well with Suite elements.

Before discussing specific curricula, however, I briefly discuss what I mean by a "research-based curriculum," describe the populations for which these curricula have been developed, and consider some of the specific goals that are being addressed. After this preamble, I briefly list the curricular materials of the Physics Suite and a few others that have been developed that match well with the Suite. In the next three chapters, I discuss these materials in detail.

RESEARCH-BASED CURRICULA

Most of the curricula that have been developed over the past few years in the United States are based at least in part on a model of student thinking and learning[1] similar to the one described in chapter 2 and have evolved using the cyclic model of curriculum development that I refer to as the research-redesign wheel. In this process, shown schematically in Figure 6.1, research on student understanding illuminates the difficulties in current instruction. The results of the research can be used to design new curricula and teaching approaches that lead to modified instruction. Research and evaluation informs on the state of effectiveness of the instruction and illuminates difficulties that remain. This process begins again and cycles in a helix of continuous educational improvement.

[1] In some, the dependence on a model of thinking and learning is tacit.

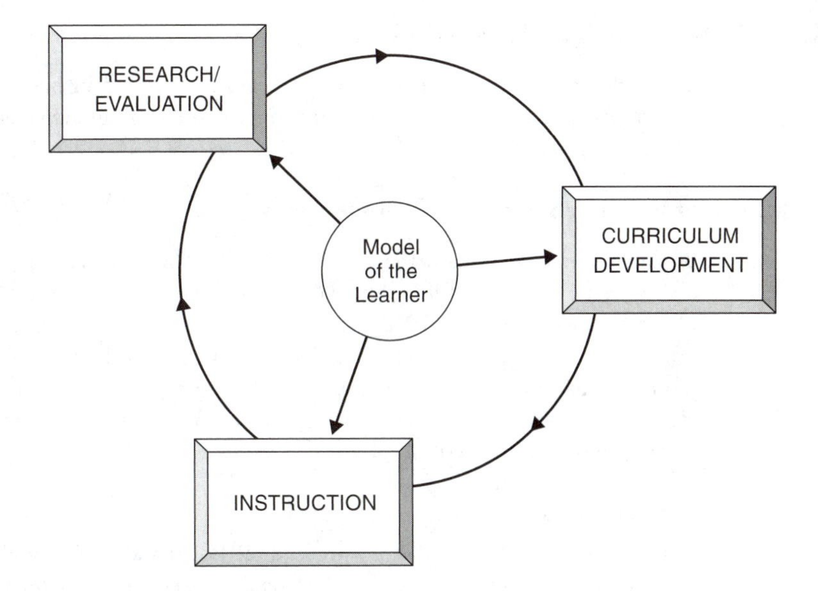

Figure 6.1 The Research and Redesign Wheel—the role of research in curriculum reform.

Of course, to understand what one sees in a research situation, one must have a model or theory of the system under investigation in order to know what to look for and to make sense of what one sees. On the other side, the experimental observations may cause us to refine or modify our theoretical model. So to the wheel, I add an *axle*—with the model of cognition and learning serving as the point about which the wheel rotates.

The research and evaluation components in this model lead to a cumulative improvement of the curriculum that is usually absent when individual faculty members develop materials in response to local needs. In the three decades that I have been a faculty member at the University of Maryland, I have watched my colleagues, highly intelligent, dedicated to their educational tasks, and concerned about the students' lack of learning in the laboratory, modifying and redesigning the laboratories for populations of students ranging from preservice teachers to engineers and physics majors. Each faculty member changes something he or she finds ineffective and makes what he or she thinks is an improvement. But since the purpose of the change is not shared, since the value of the change is not documented, and since the culture of instruction tends to focus on the individual instructor's perception of what is good instruction, the next instructor is likely to undo whatever changes have been made and make new changes. Instead of a cumulative improvement, the curriculum undergoes a drunkard's-walk oscillation.[2] The addition of the research/evaluation component to the cycle and the input from our theoretical understandings of the student and of the learning process enable us to produce curricula that can be considerably more effective than those produced by individual faculty working alone.

[2] Perhaps one can expect a long-term improvement but only proportional to the square root of the time!

MODELS OF THE CLASSROOM

Most physics instruction in the United States is delivered in one of two kinds of environments: the traditional, instructor-centered structure, and an active-engagement student-centered structure.

The traditional instructor-centered environment

If the class is large (>50), there are usually three hours of class per week, with all the students meeting together. Often, there is a weekly two- or three-hour laboratory associated with the class, uncoordinated with the lecture. If there is sufficient staff (such as graduate students to serve as teaching assistants), there may be one or two hours a week of recitation—a session in which the class is divided into small groups (<30). This traditional model of introductory physics has a number of characteristics. As taught in the United States it has the following common features:

- It is content oriented.
- If there is a laboratory, it is two to three hours and "cookbook" in nature; that is, students will go through a prescribed series of steps in order to demonstrate the truth of something taught in lecture or read in the book.
- The instructor is active during the class session, while the students are passive (at least during lectures and often during recitation).
- The instructor expects the students to undergo active learning activities on their own outside of the class section, in reading, problem solving, etc., but little or no feedback or guidance is given to help the students with these activities.

For most students, the focus of the class is the lecture. The nature of this experience can be seen clearly in the structure of the classroom. A typical lecture room is illustrated in Figure 6.2. All students are turned to face the lecturer—the focus of all attention. There may be a strong tendency for the instructor to do all the talking and to discourage (or even to suppress) student questions or comments.

The active-engagement student-centered environment

An *active-engagement* class has somewhat different characteristics.

- The course is *student centered.* What the students are actually doing in class is the focus of the course.
- Laboratories in this model are of the *guided discovery* type; that is, students are guided to observe phenomena and build for themselves the fundamental ideas via observation.
- The course may include explicit training of reasoning.
- Students are expected to be intellectually active during the class.

Active-engagement classes may occur as part of a larger class—as a recitation or laboratory combined with a traditional lecture. The smaller units have a classroom structure that

Figure 6.2 A typical lecture classroom. Even when the lecturer is superb, the focus of the activity tends to be on the lecturer, not the students. (Here, Jim Gates presents one of his popular public lectures on string theory. Courtesy Dept. of Physics, Univ. of Maryland.)

looks something like Figure 6.3. Students' attention is focused on their work and on their interaction with the other students in their group. *Facilitators* roam the room while the students are working, checking the students' progress and asking guiding questions. There may be one or more facilitators, and they may be faculty, graduate assistants, undergraduates who have had the class previously, or volunteers looking to gain teaching experience.

I refer to such an arrangement as an *active-engagement classroom.* Of course, the structure of the room does not guarantee what will happen in that room. You can do a mindless

Figure 6.3 The arrangement of an active-engagement classroom [Steinberg 2001].

cookbook lab in one of these classrooms just as easily as a highly effective discovery lab. But the structure of the room does constrain the possibilities. You can do activities in this kind of room that would be extremely difficult to carry out in a large lecture hall.

A specific type of active-engagement classroom is the *workshop* or *studio* class. In this environment, the lecture, laboratory, and recitation are combined in a single classroom. In workshop classes, most of the class time is taken up by periods in which the students are actively engaged in exploring the physics using some laboratory equipment, often involving computers in order to allow efficient high-quality data collection and to provide computer modeling tools. Only a small fraction of the period may be spent with a teacher lecturing to the students. One example of a workshop classroom is the interesting layout developed for Workshop Physics by Priscilla Laws and her collaborators at Dickinson College (see Figure 6.4). Students work two per computer station at tables with two stations. The tables are shaped so that neighboring pairs can easily collaborate. The room is set up so that there is a group interaction space in the center where demonstrations can be carried out and where the teacher can stand and easily view what is on every computer screen. This feature has the great advantage of helping the instructor identify students who might be in trouble or not on task. There is a table with a screen and blackboard at one end so that the instructor can model problem solving, do derivations, or display simulations or videos. The materials developed for Workshop Physics are a part of the Physics Suite and are discussed in detail in chapter 9.

Figure 6.4 A typical workshop or studio classroom layout (Courtesy Kerry Browne, Dickinson College).

Other arrangements for workshop-style classes have been developed at RPI for Studio Physics and at North Carolina State for the SCALE-UP project. The SCALE-UP project is discussed as a case study for the adoption and adaptation of Suite materials in chapter 10.

There is evidence that active-engagement characteristics alone do not suffice to produce significant gains in student learning [Cummings 1999]. The presentation of traditional materials in an active-engagement learning environment does not necessarily result in better concept learning than a traditional environment. What seems to be necessary is that *specific attention is paid to the knowledge and beliefs students bring into the class from their experience and previous instruction.*

THE POPULATION CONSIDERED: CALCULUS-BASED PHYSICS

As of this writing, the process of research-based curriculum development is farthest along for the introductory calculus-based ("university") physics course and the course taken by preservice elementary school teachers. The Physics Suite primarily addresses the former group (though Physics by Inquiry and Explorations in Physics specifically address the latter).

Characteristics of calculus-based physics students

Calculus-based (university) and algebra-based (college) physics courses are usually the largest service courses presently offered by physics departments.[3] At present, most of the curricular materials that have been developed have been created with the calculus-based physics class in mind. The students in this class have a number of characteristics that distinguish them from other students.

- They are mostly mathematically relatively sophisticated.
- Almost all have studied physics in high school and done well in it.
- Almost all think physics is important for their careers.
- They mostly consider themselves scientists or engineers.

The hidden curriculum and problem solving

When we talk about our classes, we usually specify a certain set of content. But if all our students take away from our course is content, their ability to use this content may be limited. They may have developed a vocabulary, learned to recognize that they've seen a particular equation before, and may perhaps have improved their algebraic skills somewhat. This is not enough to keep students taking these courses at a time when there is great competition for places in the engineering curriculum, and it does not scratch the surface of the powerful and valuable skills and attitudes that could be delivered. I refer to the (usually tacit) gains that we hope our students will achieve as a result of taking a physics course as the *hidden curriculum.*

[3] There are some tantalizing counterexamples that illustrate possibilities for future developments. One example is Lou Bloomfield's class for nonscientists at the University of Virginia, "How Things Work," using his text of the same name [Bloomfield 2001]. As of this writing, I am unaware of any research on the results of this class on student learning or understanding.

We began to discuss the hidden curriculum in chapter 3. Here, let's try to explicate some of those elements that might be important for developing authentic problem-solving skills, based on the understanding of student learning we have developed in previous chapters.

The research on problem solving shows that experts use a good understanding of the concepts involved to decide what physics to use. Novices look for an equation. Experts classify problems by what physics principles are most relevant, such as energy vs. force analysis. Novices classify them by surface structure and superficial associations (e.g., it's an inclined plane), and they remember a particular problem they did with inclined planes [Chi 1981]. We would really like our students to learn the components of problem solving used by expert physicists:

- The ability to "find what physics will be useful" for a problem
- The skill to take apart and solve complex problems
- The ability to evaluate the result of a solution and know whether it makes sense

In order to achieve all of these goals, a student has to be able to make sense of what a problem "is about." In order to develop such a mental model, an understanding of the concepts—of the physical meaning of the terms and symbols used in physics—is essential (necessary, but not sufficient). As described in chapter 1, success in algorithmic problem solving has been shown to be poorly correlated with a good understanding of basic concepts [Mazur 1997] [McDermott 1999]. This observation fits well with the cognitive structures described in chapters 2 and 3.

SOME ACTIVE-ENGAGEMENT STUDENT-CENTERED CURRICULA

The physical (and temporal) architecture of the classroom is only one part of what controls what happens to students; the other is the cognitive architecture, which is determined by the curricular materials and by how the instructor uses them. The curricula associated with the Physics Suite, and the additional curricula I have chosen to include, are coherent, rely on the educational principles discussed in the first half of this book, and focus on getting students "to do what needs to be done." Most of them focus on the goal of improving student conceptual understanding and their ability to use these concepts in complex problem solving.

Models of instruction have been developed that replace one or more of the elements of the traditional structure by an active-engagement activity. *Lecture-based models* modify the traditional lecturer presentation to include some explicit student interaction. *Laboratory-based models* replace the traditional laboratory by a discovery-type laboratory. *Recitation-based models* replace the recitation in which an instructor models problem solving for an hour by a structure in which the students learn reasoning or problem solving in groups guided by worksheets. They may also carry out qualitative guided-discovery experiments. Finally, some models go beyond the traditional structure by creating an environment that combines elements of lecture, laboratory, and recitation in a single class, usually dominated by guided-discovery laboratories. I refer to these as *workshop models.*

Following are the models that I discuss in the next few chapters. The specific materials that have explicitly been coordinated as part of the Physics Suite are marked in bold.

Lecture-based models (chapter 7)

- Traditional lecture
- Peer Instruction/ConcepTests
- **Interactive Lecture Demonstrations**
- Just-in Time Teaching

Recitation-based models (chapter 8)

- Traditional recitation
- **Tutorials in Introductory Physics**
- **ABP Tutorials**
- Cooperative Problem Solving

Laboratory-based models (chapter 8)

- Traditional laboratory
- **RealTime Physics**[4]

Workshop models (chapter 9)

- **Physics by Inquiry**
- **Workshop Physics**
- **Explorations in Physics** (not discussed in this volume)

In the next three chapters, the discussion of each model begins with a boxed summary; each summary describes briefly the following elements:

- The ***environment*** in which the method is carried out (lecture, lab, recitation, or workshop)
- The ***staff*** required to implement the method
- The ***populations*** for which the method has been developed and tested and those to whom it might be appropriately extended
- Whether ***computers*** are required to implement the method and how many
- Other ***specialized equipment*** that might be required
- The ***time investment*** needed to prepare and implement the method
- The ***materials and support*** that are available

Within the description of the method itself, I discuss the method briefly, consider some explicit example, and, if there is data on the method's effectiveness, I present some sample data. If I have had personal experience with the method, I discuss it.

[4] *Tools for Scientific Thinking*, a somewhat lower level set of laboratory materials similar in spirit to *RealTime Physics*, are also a part of the Physics Suite but are not discussed in this volume.

CHAPTER 7

Lecture-Based Methods

When I, sitting, heard the astronomer,
where he lectured with such applause in the lecture room,
How soon, unaccountable, I became tired and sick;
Till rising and gliding out, I wander'd off by myself,
In the mystical moist night-air, and from time to time,
Look'd up in perfect silence at the stars.
Walt Whitman

Most of the introductory physics classes in the United States rely heavily on the traditional lecture. Research has rather broadly shown (see, e.g. [Thornton 1990]) that lectures, even when given by good lecturers, have limited success in helping students make sense of the physics they are learning. Good lectures can certainly help motivate students, though, as I discuss in chapter 3, lecturers often don't know how to help students convert that motivation to solid learning.

Even in a traditional class with a large number of students, there are some things you can do to get your students more engaged during a lecture. Unfortunately, some of the "obvious" things that both Sagredo and I have tried to do in lecture—such as asking rhetorical questions, asking them to think about something I've said, telling them to make a prediction before a demonstration (but not making those predictions public), having them work out something in their notebooks, or even doing lots of demonstrations—don't seem to have much effect. Something more structured seems to be required—something that involves having them give explicit responses that are collected and paid attention to.

In this chapter, I discuss my experience in traditional lectures and provide some detailed tips that in my experience can help improve that environment. Then I describe three models that involve more structured interactions with the students and that have been shown to produce dramatic improvements in student learning: Peer Instruction, Interactive Lecture Demonstrations, and Just-in-Time Teaching.

THE TRADITIONAL LECTURE

Environment: Lecture.
Staff: One lecturer (N = 20–600).
Population: Introductory algebra- or calculus-based physics students.
Computers: None required.
Other Equipment: Traditional demonstration equipment.
Time Investment: 10–20 hours planning time per semester, 1–2 hours preparation time per lecture.

Traditional lectures offer opportunities to inspire and motivate students, but one shouldn't make the mistake of assuming that students immediately understand and learn whatever the professor says and puts on the board. I consider myself a good "learner-from-lectures," having had years of experience with them. I regularly attended my lectures in university and grad school, took excellent notes, and studied from them. As a researcher, first in nuclear physics and now in physics education, I attend dozens of seminars and conference lectures every year. I enjoy them and feel that I learn from them.

But occasionally, I've been brought up short and have been reminded what the experience is like for a student. I still vividly recall a colloquium given some years ago at the University of Maryland by Nobel Laureate C. N. (Frank) Yang on the subject of magnetic monopoles. This was a subject I had looked at briefly while in graduate school, had some interest in, but hadn't pursued at a professional level. I had all the prerequisites (though some were rusty) and was familiar with the issues. Yang gave a beautiful lecture—clear, concise, and to the point. I listened with great pleasure, feeling that I finally understood what the issues about magnetic monopoles were and how they were resolved. Leaving the lecture, I ran into one of my friends and colleagues walking toward me. "Oh, Sagredo," I said. "You just missed the greatest talk!"

"What did you learn?" he asked.

I stopped, thought, and tried to bring back what I had just heard and seen in the lecture. All I could recall was the emotional sense of clarity and understanding—but none of the specifics. I was left with the only possible response, "Frank Yang really understands monopoles."

Once I had my grad assistants (ones not associated with the course) stationed waiting at the top of my lecture hall after lecture, grabbing students leaving at the end of the class and asking them "What did he talk about today?" Of the students willing to stop and chat, almost none could recall anything about the lecture other than the general topic.

This could in principle be an acceptable situation. If I had taken good lecture notes in Yang's lecture, I could have gone back to look at them and spent the time weaving the new information into my existing schemas. (I thought I had "listened for understanding" instead.) Unfortunately, many of my students do not take good lecture notes. Of those who do, many do not know how to use them as a study aid. Of those who know how to use their notes, many are highly pressed for time by other classes, social activities, or jobs, and can't devote the time required for the task. The assumption that "most students use lectures to create good lecture notes which they then study from" can be a very bad assumption indeed.

A more interactive approach to the traditional lecture

An alternative approach is to use the lecture in a more interactive way. Even within the framework of the traditional lecture, there are many tricks the instructor may use to increase the student's intellectual engagement in the class.[1] Some are fairly obvious and are taught in classes for new faculty, probed in end-of-the-semester student questionnaires, and watched for by peer evaluators. They include:

- Speak clearly and at an appropriate pace.
- Write on the board using good handwriting and good layout.
- Give students sufficient time to copy anything you expect them to copy.

Lecturers are often not aware of defects along these lines, focusing on the content rather than on what they are saying about the content. Problems with these issues can be helped by videotaping and reviewing your lectures or by having a sympathetic colleague sit in the back and watch your presentation.

Some of the things you can do to keep the students interested and attentive are the following:

- Set the context.
- Chunk the material.
- Facilitate note-taking.
- Develop a good speaking technique.
- Ask authentic questions.
- In discussions, value process as well as right answers.
- Get students to vote on a choice of answers.
- Make it personal.

Set the context

I once heard Sagredo deliver a lecture to a graduate class of physics students on a subject I thought I needed to learn more about. The lecture was enlightening—and I learned what I needed to know—but he presented it in a way I found disturbing. He began with 45 minutes of technical development without any discussion of his motivation or why this development was going to be interesting or useful. In the last 5 minutes, he wrapped everything together in an elegant package, applying all the technical details to the case of interest. When I asked him why he approached the lecture in this way, he said, "I didn't want to give away the punch line." Sagredo, I think stand-up comedy is the wrong metaphor for a physics

[1] Many of these and more are discussed in Donald Bligh's useful book, *What's the Use of Lectures?* [Bligh 1998].

lecture. Although our students seem to be accustomed to trying to take in random, unmotivated mathematical results, I don't think that is the best way to engage their attention and interest.

Since everyone's thinking and learning is naturally associative, we can expect to get the best results by tying new material to something the student already knows.[2] I try to begin my lectures by setting a context, letting the students know beforehand what the point of the lecture is and where we are going. Before every lecture, I write an outline of what we will be doing on the upper-left corner of the board so that students can have some idea of what we are going to be talking about.

Chunk the material

Another thing I try to do when lecturing in a large hall with many students is to keep in mind the difficulty produced by the limits to working memory. You cannot expect your students to keep a large number of difficult ideas in mind for a long time and bring them together at the end as you tie everything up into a neat package.

I try to chunk my lectures into coherent pieces that begin at the upper left and that can be completed on the available board space. Once I've finished the chunk, I don't just continue, but I stop, walk to the back of the class, wait until the students finish their note-taking, and go over the entire argument again so that the students have a context and can see the entire presentation at once. Summarizing after the chunk is complete helps students find a way to integrate the new material with their existing knowledge structures.

Facilitate note-taking

Tricks to speed things up, such as using pre-prepared transparencies, are usually counterproductive, especially if you are expecting students to take notes. Copying something from the board usually takes more time than it does to write it on the board, since the copyist also has to read and interpret what's been written. Forcing yourself to write it on the board at least gives you some idea of what the students are going through. Even if you do not expect the students to copy from a display, you are likely to severely underestimate the time it takes the students to read and make sense of the material you put up since you are very familiar with it and they are not.

Handing out previously prepared lecture notes may in principle help a bit, but since the instructor (rather than the student) is preparing the notes, the instructor gains the associated processing benefits, not the student. A more plausible approach[3] is to provide students with a set of "skeleton" notes—with just the main points sketched out but with spaces in which the students are supposed to fill-in what happens in class. This could be useful in helping students to create a well-organized set of notes and might help them in following the lecture. I have adapted this idea by creating PowerPoint presentations for each lecture that have a similar "skeleton" structure. The students can choose to print these out as handouts. I do derivations and problem solutions on the board so they don't go by too quickly—but the figures and diagrams can be more neatly prepared on the computer.

[2] Recall the example with the strings of numbers in chapter 2.

[3] I learned this trick from Robert Brown at Case-Western Reserve University.

Develop a good speaking technique

Getting students in a large class to engage the material is most easily accomplished by having them engage in carefully designed individual and group activities. The research-based curricular materials discussed below give some examples of how this can be done. Even without the use of prearranged materials, you can improve your students' engagement somewhat in these ways:

- *Talk to the students*—Face the students when you speak. If you have to write something on the board, do not "talk to the board" with your back to the students. Write and then turn to the class to describe or explain it.
- *Use appropriate tones of voice*—Learn to project your voice. Test your classroom with a friend, seeing if you can be heard adequately from all parts of the room. If you can't, use a microphone. Be careful! It's natural to project in a loud voice during the test and to forget to do it when you get involved in what you are saying in lecture. One trick that seems effective is, after announcing that something is important, drop your voice a bit to present the important information. The class will quiet significantly to hear what you are saying.
- *Step out of the frame*—In a lecture hall, walk up the aisles and speak from the middle or the back of the class. This requires the students to turn around. Changing their orientation restores their attention (at least momentarily—and it allows you to stand next to and stare down a student who is reading a newspaper). This also breaks the imaginary "pane of glass" the students put up between you and them and helps to change you from a "talking head on a TV screen" (to whom they feel no need to be polite or considerate) to a human being (to whom they do).
- *Make eye contact*—When you look a student in the eye during your lecture, for that student, you change the character of the activity from a TV- or movie-like experience into one more like a conversation, even if it's only for a moment, and even if you are doing most of (or all of) the talking. But be careful not to fixate on one particular student. That can be intimidating for that student. Switch your gaze from student to student every few seconds.

Ask authentic questions

An excellent way to get students involved is to ask questions to which they respond by really thinking about and answering them. This can be very effective, but it is harder than it sounds. Most faculty questions are rhetorical—that is, they are not meant to be answered by the students—in practice, if not in intent. Faculty tend to be as nervous about "dead air" as a TV news anchor. Two or three seconds of silence can seem like an eternity while you are waiting for students to answer a question you've posed. The easiest solution is to answer it yourself. But students know that faculty do this, so they wait you out. To get them to realize that you really do want them to answer and that the question is not just a part of the lecture, you have to outwait them—at least until they get in the habit of answering. This can be quite painful until you get used to it. The idea is to wait until they get uncomfortable with the silence, and this can easily take 20 to 30 seconds or longer. Sometimes you may have to reiterate your question or call on a specific student at random to show you really want a response.

You also have to build your students' confidence that answering questions is not going to be a painful experience. Students are most reluctant to look foolish in front of their instructor (and in front of their peers), so it is often quite difficult to elicit responses to questions. Being negative or putting down a student's question or answer can in one sentence establish a lack of trust between the instructor and the class that can last for the rest of the term. The result can be a class in which the instructor does all the talking, severely reducing the students' involvement and attention. I feel strongly enough about this principle that I set it off as

> *Redish's seventh teaching commandment:* Never, ever put down a student's comment in class or embarrass a student in front of classmates.

This isn't always easy, even for a supportive and compassionate lecturer. I remember one occasion some years ago in which I was lecturing to a class of about 20 sophomore physics majors on the Bohr model. As I proceeded to put down a blizzard of equations (laid out most clearly and coherently, I was certain), one student stopped me and asked: "Professor Redish, how did you get from line 3 to line 4?" I looked at the equations carefully and mentally reminded myself that the student asking the question was mathematically quite sophisticated and was taking our math course in complex variables that semester. After a pause during which I suppressed a number of put-downs and nasty remarks, I responded: "You multiply both sides in line 3 by a factor of two," before proceeding without further comment. Thinking about the situation later, I realized that I had been going too fast, using too many equations without sufficient explanation, and not giving the students enough time to follow the argument.

In discussions, value process as well as right answers

Another important step in being able to build a class that responds and participates in discussion is to change the class's idea that you are looking for the "right answer." If that's all you ever ask for, only the few brightest and most aggressive students will answer your questions. This will only reconfirm the attitude that most students have that science in general and physics in particular is a collection of facts to be memorized rather than a process of reasoning—and that only a few really bright people can do it. (See chapter 3.)

Even if the first student answering your question gives the correct answer, one way to begin to break this epistemological misconception is to ask for other possible answers, emphasizing creativity and explaining that the students are not required to believe the answers they give. I'll give an example of my experience with this technique in the section under Interactive Lecture Demonstrations.

Get students to vote on a choice of answers

Even if only a small number of students are willing to respond to an instructor's question, there are still ways of engaging a larger fraction of students in a lecture. One of the easiest and most effective is voting. This is easy to implement. Set out some options and ask the class to raise their hands in support of the different options. In some classes, only a few students will respond. If the voting process is to work to keep them engaged, this has to be overcome. In such a case, I often walk up the aisle and point to someone who hasn't voted and ask them to explain their difficulty and why they were unable to make a decision.

Another idea is to give each student a set of five "flash cards" at the beginning of the term.[4] These should be large (the size of a notebook page) and have one of the letters "A" through "E" on one side, and other options ("true," "false," "yes," "no," and "maybe" or "?") on the back. It may help to make them different colors. The students are instructed to bring their flashcards to every lecture. (If you really want them to do this, you have to use them at least once in every lecture, preferably more often than that.) When you present your choices, you label them with the letters or other options, and the students hold up the answer they choose. You can easily see the distribution of answers. If the flashcards are not colored, it's important to report back to the students an approximate distribution of the voting. If you use colored cards, they will be able to see each other's cards. The sense of not being alone in their opinions is an important part of increasing their comfort with taking a stand that might turn out to be wrong. There are electronic versions of this system available in which each student gets a remote-control device on which they can click their answers. These answers are beamed to a collector and displayed on a computer projection screen.

Make it personal

Finally, perhaps the most important component of delivering effective lectures (and classes in general) is to show the students that you are on their side.

> *Redish's eighth teaching commandment*: Convince your students that you care about their learning and believe that they all can learn what you have to teach.

This can make a tremendous difference in a class's attitude, no matter what environment you are using. One way to demonstrate this caring is to learn as many of the students' names as you can. Even if you only learn the names of the students who ask questions in class, it will give the rest of the students the impression that you know all (or most) of them. I take photographs in recitation section and copy them. (This also helps my teaching assistants learn the students' names.) I then bring them with me to class and spend three to five minutes before class matching names to faces. After class, I check any students who have come up to ask questions. It turns out to be relatively easy to learn the names of 50 to 100 students without much effort. Not only do the students get more personally engaged in the class, but so do I.

Demonstrations

An important component of a traditional introductory physics lecture is the lecture demonstration. Sagredo suggested to me that perhaps he should just do more demonstrations "especially since they don't seem to follow the math very well. After all, seeing is believing." I wish it were that easy, Sagredo. As physicists, we are particularly enamored of a good demonstration. After all, if they are properly set up, a demonstration makes clear what the physics is—doesn't it?

[4] I learned this idea from Tom Moore. See [Moore 1998] and [Meltzer 1996].

Unfortunately, demonstrations are not always as effective as we expect them to be, for two reasons coming from our cognitive model:

- Students may not see demonstrations as important.
- Students may not see in a demonstration what we expect them to see.

Sagredo sat in on some of my lectures in the large calculus-based class for engineers to help me evaluate my presentation and to consider ideas that might be adaptable to his own class. At one point in one of the lectures, I did a demonstration. The equipment had been prepared by our superb lecture demonstration facility and was large and visible throughout the hall. Furthermore, it worked smoothly. A few students asked questions at the end. Overall, I was pleased and felt it went well.

Sagredo came up to me after class. "You will never guess what happened in your demonstration!" he said. "Fully half the class simply stopped paying attention when you brought out the equipment! Only the group in the front few rows and a few scattered around were really trying to follow. Lots of students pulled out their newspapers or started talking discretely to friends!"

Since I had been concentrating on the equipment—and on the students in the first few rows—I hadn't noticed this. It's plausible, though. As discussed in chapter 3, students' expectations about the nature of learning and their goals for the class play a big role in filtering what they will pay attention to in class. At the time, it was not my habit to ask exam questions about demonstrations, so they were reasonably certain they wouldn't be asked about it.

My next step, then, was to change how I did my demonstrations. I began to do them less frequently but to spend more time on each one. I tried to engage more of the class and assured them that there would be an exam question on one of our demonstrations. In this environment, I learned something even more striking—but not surprising given our cognitive model.

Physics education researchers learned many years ago that students often think that circular motion tends to persist after the forces producing it are removed [McCloskey 1983]. In order to attempt to deal with the common naïve conception, I did the demonstration shown in Figure 7.1. A circular ring about 0.5 m in diameter with a piece cut out of it (about

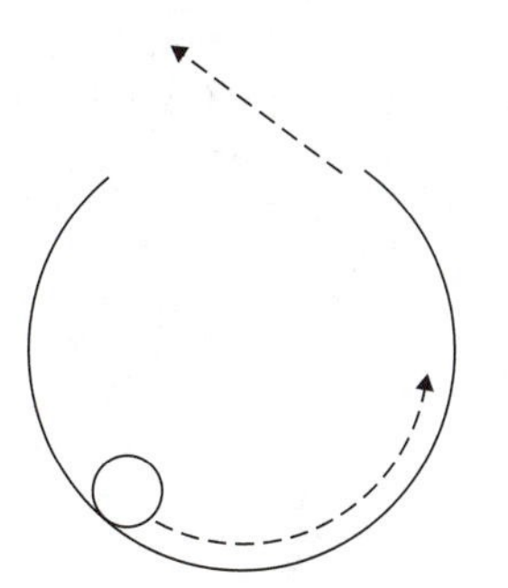

Figure 7.1 Lecture demonstration on circular motion (after [Arons 1990]). A partial circular ring lies flat on a table, and a billiard ball is rolled around the ring.

60° worth) was laid flat on the table. The point of the demonstration was to show that a billiard ball rolled around the ring would continue on in a straight line when it reached the end of the ring.

I did the demonstration in the following series of steps in order to engage more of the students' attentions in what was happening.

1. I briefly reviewed the physics—circular motion and Newton's second law.
2. I showed the apparatus and showed what I was going to do. I rolled the ball along the ring but stopped it before it got to the edge.
3. I asked students what they expected would happen. Some expected the correct straight line, but most expected it would continue to curve a bit. I called for discussion, and a number of students defended one answer or another.
4. I put the answers on the board and asked for a show of hands. It was split, with a substantial number of students supporting each answer. (No one thought it would continue on in the circle when there wasn't any ring holding it in.)
5. I then showed the demonstration, letting the ball roll on beyond the edge of the ring.

Then, by a lucky chance, instead of saying: "There. You see it goes in a straight line," I asked them what *they* saw. To my absolute amazement, nearly half the students claimed that the ball had followed the curved path they expected! The other half argued that it looked to them like a straight line. Lots of mini-arguments broke out among the students. Somewhat nonplussed, I looked around and found a meter stick. "Let's see if we can decide this by looking a bit more carefully. I'll align the meter stick along what a straight path would be—tangent to the point where it will leave the circle—and about an inch away. If it's going straight, it will stay the same distance from the ruler. If it curves, it will get farther away from the ruler as it goes." Now, when I did it, the path was obviously straight since it remained parallel to the ruler. It was only at this point that I got the gasp I had expected from half the class.

Now that I know "what they need," I could do the demo in the future using the ruler right away. But I feel that would be a mistake. The predictions and discussions, the taking a stand and defending their point of view, the surprise at having mis-seen what was happening—all of these contribute to the students' engagement in and attention to the activity. Although I have no hard comparative data (it would make a nice experiment), I expect the demo we did was much better remembered than if I had simply "done it right." On the midsemester exam, I gave the relevant question from the Force Concept Inventory, and more than 80% of the students gave the correct response. This is much better than the typical results from traditional instruction.

PEER INSTRUCTION/CONCEPTESTS

Environment: Lecture.

Staff: One lecturer trained in the approach ($N = 30–300$).

Population: Introductory algebra- or calculus-based physics students (though some ConcepTest questions are appropriate for less sophisticated populations).

Computers: None required, but one associated with a response system permits live-time display of quiz results.

Other Equipment: Some kind of student-response system. This can be as low tech as cards for the students to hold up or as high tech as a computer-based system with individual wireless remote-response devices for each student.

Time Investment: Low to moderate.

Available Materials: Text with ConcepTest questions [Mazur 1997]; http://galileo.harvard.edu

Eric Mazur describes his method for increasing students' engagement in his lectures in his book *Peer Instruction* [Mazur 1997]. His method includes three parts:

1. A web-based reading assignment at the beginning of the class (see the section on JiTT below)
2. ConcepTests during the lecture
3. Conceptual exam questions

During the lecture he stops after a five- to seven-minute segment to present a challenging multiple-choice question about the material just covered (a *ConcepTest*). This question is concept oriented, and the distractors are based on the most common student difficulties as shown by research. Students answer the questions at their seats by either holding up a colored card showing their answer or by using a device that collects and displays the collective response on a projection screen, such as *ClassTalk*™ or the *Personal Response System*™.

Mazur then instructs the students to discuss the problem with their neighbor for two minutes. At the end of this period, the students answer the question again. Usually the discussion has produced a substantial improvement. If not, Mazur presents additional material. A sample of one of Mazur's questions is given at the top of Figure 7.2. The ConcepTest discussion takes another five to seven minutes, breaking the lecture up into 10- to 15-minute chunks.

The response of Mazur's students in a Harvard algebra-based class to this question is shown in the lower half of Figure 7.2. Note that about 50% of the students start with a correct answer before discussion and about 70% have the right answer after discussion. What's more, the fraction of students who have the right answer and are confident about it increases from 12% to 47%. This is a rather substantial learning gain for two minutes of discussion time.

Mazur suggests that a question used in this way should be adjusted so that the initial percentage correct is between 35 and 70%. Less than this, and there will be too few students with the correct answer to help the others. More than that, and either you haven't found the right distractors or enough students know the answer that the discussion isn't worth the class time.

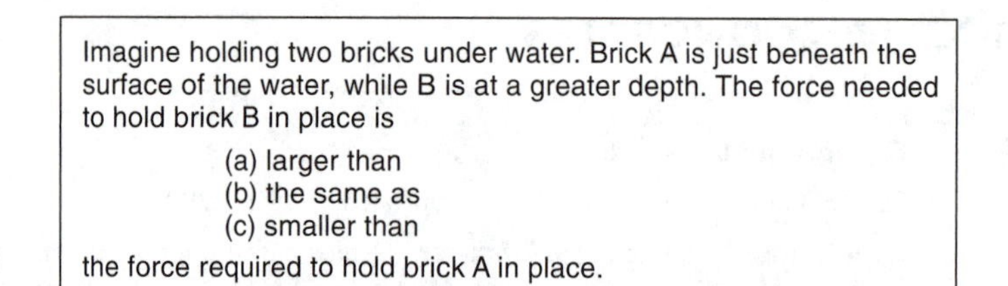

Imagine holding two bricks under water. Brick A is just beneath the surface of the water, while B is at a greater depth. The force needed to hold brick B in place is

(a) larger than
(b) the same as
(c) smaller than

the force required to hold brick A in place.

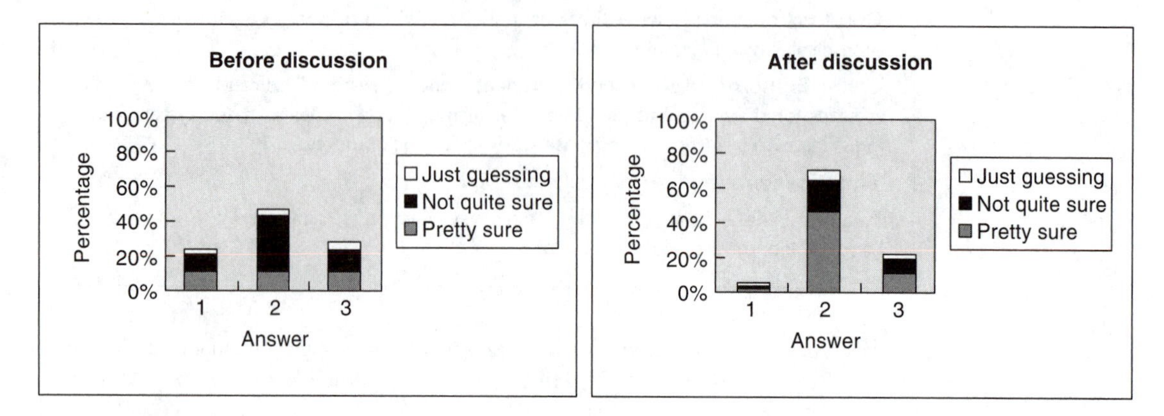

Figure 7.2 A ConcepTest question with the results before and after discussion (from [Mazur 1997]).

For all the ConcepTest questions in a given semester, Mazur found that the fraction of correct answers invariably increased after the two-minute discussions. A plot of this result is shown in Figure 7.3.

Finally, sensitive to the principle that students only focus on things that you test, Mazur includes conceptual questions on every exam. His book contains reading quizzes, ConcepTests, and conceptual exam questions for most topics in the traditional introductory physics course.

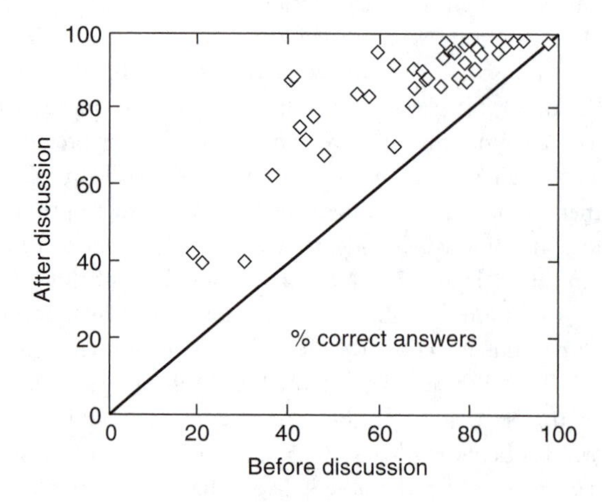

Figure 7.3 The fraction of correct answer before and after peer discussion (from [Mazur 1997]).

INTERACTIVE LECTURE DEMONSTRATIONS (ILDS)

Environment: Lecture.

Staff: One lecturer trained in the approach (N = 30–300).

Population: Introductory algebra- or calculus-based physics students (though some ILDs are appropriate for less sophisticated populations).

Computers: One required for the lecturer.

Other Equipment: LCD or other appropriately large-screen display. Some kind of computer-assisted data acquisition device. Specific standard lecture-demonstration equipment is required for each ILD.

Time Investment: Low.

Available Materials: Worksheets for about 25 ILDs [Sokoloff 2001].

An approach that has proven both effective and efficient is a series of interactive lecture demonstrations (ILDs) by Sokoloff and Thornton [Sokoloff 1997] [Sokoloff 2001]. These demonstrations focus on fundamental conceptual issues and take up a few lecture periods (perhaps four to six) during a semester. Most use computer-assisted data acquisition to quickly collect and display high-quality data.

In order to get the students actively engaged, each student is given two copies of a worksheet to fill out during the ILD—one for predictions (which they hand in at the end) and one for results (which they keep). Giving a few points for doing the ILD and handing in the prediction sheet (which should *not* be graded) is a valuable way to both increase attendance and get some feedback on where your students are.[5]

Each ILD sequence goes through a series of demonstrations illustrating simple fundamental principles. For example, in the kinematics demonstrations, demonstrations are carried out for situations involving both constant velocity and constant acceleration. The cases with acceleration use a fan cart to provide an approximately constant acceleration. (See Figure 7.4.) The demonstrations address a number of specific naïve conceptions, including confusion about signs and confusion between a function and its derivative. A fan is used to provide an acceleration rather than gravity so as not to bring in the additional confusion caused by going to two dimensions. The specific demonstrations are:

1. Cart moving away from motion detector at constant velocity
2. Cart moving toward the motion detector at a constant velocity
3. Cart moving away from the motion detector and speeding up at a steady rate
4. Cart moving away from the motion detector and slowing down at a steady rate (fan opposes push)

[5] Sokoloff reports that even though students are told that the prediction sheets will not be graded and that they should leave their original predictions to be handed in, some students correct their prediction sheet to show the correct answer instead of their prediction.

5. Cart moving toward the motion detector and slowing down at a steady rate (fan opposes push)
6. Cart moving toward the motion detector and slowing down, then reversing direction and speeding up

In each case, the demonstrator goes through the following steps:

- Describe the demonstration to be carried out, performing it without collecting data.
- Ask the students to make and write down individual predictions on their prediction sheets ($\Delta t \sim$ one minute).
- Have the students discuss the results with their neighbors and indicate their consensus prediction on their prediction sheets ($\Delta t \sim$ two to three minutes).
- Hold a class discussion, putting the various predictions on the board.
- Perform the demonstration, collecting data and having the students copy the results on their results sheet.
- Hold a brief class discussion reflecting on why the answer obtained makes sense and the other answers have problems.

I have carried out some of these ILDs in my algebra-based physics classes. The first time I did them, I was tempted to leave out some of the demonstrations, finding them repetitious. After all, once they got demo 4, isn't demo 5 obvious? When I did this, some students came

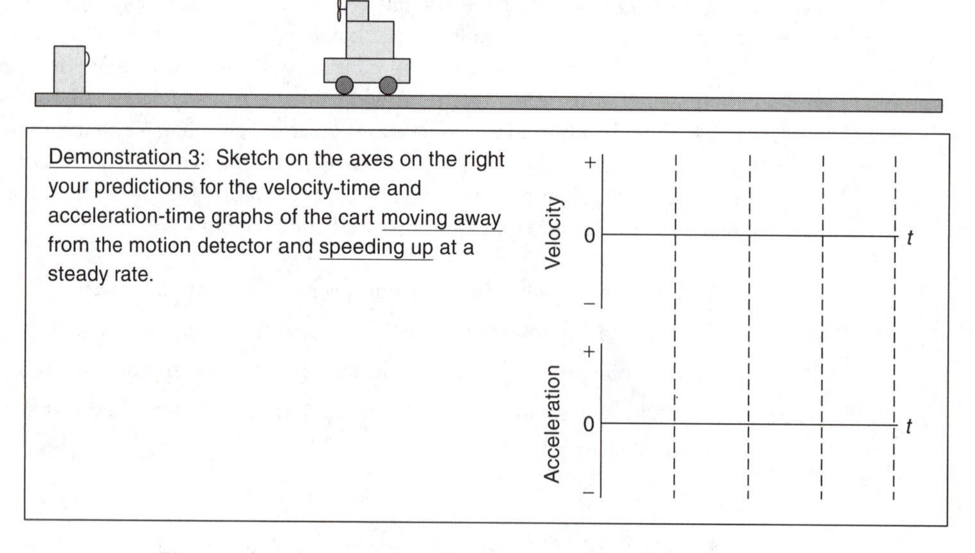

Figure 7.4 The apparatus and worksheet entry for a kinematics ILD.

up after class asking for the results. When I asked them for their predictions, they had the wrong answers, having found it difficult (not at all obvious) to make the translation from the other cases.

One place where I have found ILDs to be extremely valuable is in the class discussion step. This offers a tremendous opportunity to change the character of the class and your interaction with the students in a fundamental way. As I described in some detail in chapter 3, many of our students have the epistemological misconception that science in general, and physics in particular, is about the amassing of a set of "true facts." They think that learning scientific reasoning and sense making are a waste of time. Given this predilection, most students are reluctant to answer a question in class if they are not convinced they have the correct answer.

In discussing the predictions for the ILDs, I encourage the class to "be creative" and to find not just what they think might be the correct answer (probably the one given by the A student in the front row who answered first!), but to come up with other answers that might be considered plausible by other people (such as a roommate who is not taking the class). This frees the students from the burden of being personally associated with the answer they are giving and allows them to actually express what they might really believe. (I sometimes find it necessary to give some plausible but wrong answers myself in order to get the ball rolling.) I then ask students to try to defend each other's answers. This changes the character of the discussion from one that is looking for the right answer to one that is trying to create and evaluate a range of possible answers. The focus changes from "listing facts" to building process skills.

The results of this were quite dramatic in my class. Many more students became willing to answer (and ask) questions, and I was able to elicit responses from many more of my students than ever before. (In a class of 165 students, about 40 to 50 students were willing to participate in subsequent discussions.)

The evaluations of student conceptual improvement with ILDs were done by Thornton and Sokoloff in mechanics using the FMCE. The results they reported were spectacular, with students in classes at Tufts and Oregon improving to 70% to 90% from a starting point of less than 20%. Of course, this result is at the primary institution, and the demonstrations were performed by the developers or by colleagues they themselves have trained.

Secondary users have reported some difficulties with their implementation. One colleague of mine reported implementing ILDs and obtaining no improvement on the FCI over traditional demonstrations [Johnston 2001].[6] In my own experience with the technique, I find it not as easy to implement effectively as it appears on the surface. With traditional demonstrations, students often either sit back and expect to be entertained or tune out altogether. With ILDs, it is essential to get the students out of that mode and into a mode where they are actively engaging the issues intellectually. This is not easy, especially with a class that is accustomed to passive lecturing and instructor-oriented demonstrations. A full analysis of ILD implementation is currently under way [Wittmann 2001].

[6]In this case, the lecturer is highly dynamic and entertaining. My hypothesis is that he maintained a demonstration/listening mode in his students rather than managing to get them engaged and thinking about the problems.

JUST-IN-TIME TEACHING (JiTT)

Environment: Lecture.
Staff: One lecturer trained in the approach (N = 30–300).
Population: Introductory algebra- or calculus-based physics students.
Computers: One required for the lecturer. Students require access to the web.
Other Equipment: None.
Time Investment: Moderate to high.
Available Materials: Text with questions [Novak 1999]; website http://www.jitt.org.

The Just-in-Time Teaching or JiTT approach was developed by Gregor Novak and Andy Gavrin at Indiana University–Purdue University Indianapolis (IUPUI) and Evelyn Patterson at the U.S. Air Force Academy. The group has collaborated with Wolfgang Christian at Davidson College to create simulations that can be used over the web.

The JiTT approach is described in the group's book, *Just-in-Time Teaching: Blending Active Learning with Web Technology* [Novak 1999]. The method is a synergistic curriculum model that combines modified lectures, group-discussion problem solving, and web technology. These modifications are reasonably self-standing and can be adopted by themselves or in combination with other new methods that are described in the next two chapters.

The JiTT approach has as its goals for student learning a number of the items addressed in chapters 2 and 3:

- Improve conceptual understanding.
- Improve problem-solving skills.
- Develop critical thinking abilities.
- Build teamwork and communication skills.
- Learn to connect classroom learning with real-world experience.

To achieve these goals, JiTT focuses on two critical cognitive principles, one from each side of the teaching/learning gap:

- Students learn more effectively if they are intellectually engaged.
- Instructors teach more effectively if they understand what their students think and know.

These principles are implemented by using web technology to change students' expectations as to their role in the learning process and to create a feedback loop between student and instructor. This feedback is implemented by assigning web homework "WarmUp" assignments before each class. The components of the process are as follows.

1. Before each lecture, specific, carefully chosen *WarmUp* questions are assigned and made available on the web. The questions concern a topic that has not yet been con-

sidered in class and that will be addressed in the lecture and class discussions and activities. (The detailed character of these questions is discussed below.)

2. Students are expected to do the reading and consider the questions carefully, providing their best answers. They are graded for effort, not correctness. The student responses are due a few hours before class.
3. The instructor looks at the student responses before lecture, estimates the frequency of different responses, and selects certain responses to put on transparencies (or display electronically) to include as part of the in-class discussion and activities.
4. The class discussion and activities are built around the WarmUp questions and student responses.
5. At the end of a topic, a tricky question known as a *puzzle* is put on the web for students to answer.

The authors report that for the students, thinking about the questions beforehand, seeing their own responses as a part of class discussion, and discovering that they can solve tricky questions using what they have learned raises the level of student engagement substantially. For the instructor, the explicit display of student difficulties provides much more feedback than is typically available. This feedback can keep the instructor from assuming too much about what the students know and can help direct the class discussion to where it will do the most good.

A successful implementation of JiTT relies on:

1. *A mechanism for delivering questions over the web and for collecting and displaying student answers in a convenient form.* You can use a number of web environments such as WebAssign™, CAPA, and Beyond Question, or course management systems such as BlackBoard or WebCT.
2. *A set of carefully designed warm-up questions and puzzles that get to the heart of the physics issues.* The JiTT book includes examples of 29 threefold WarmUp assignments and 23 puzzles on the topics of mechanics, thermodynamics, E&M, and optics. Many additional JiTT materials developed by adopters and adapters are accessible via the JiTT website.
3. *An instructor with sufficient knowledge of student difficulties and with strong skills for leading a classroom discussion.* This is something that cannot be easily provided and is the reason I have rated this method as requiring a "moderate to high" time investment.

Sagredo, although I told you at the beginning of chapter 6 that I could not provide you any "best method" for teaching a particular physics topic, this approach allows you to learn about specific student difficulties and to make use of what you have learned.

Running a discussion in a large lecture in such a way that many students are involved, that the appropriate physics is covered, and that the students get to resolve their difficulties requires substantial skill. The JiTT book includes discussions of various specific examples that show the kinds of techniques that can be effective.

Since the entire structure of the class relies on the student responses to the WarmUp questions and puzzles, the choice of these questions becomes critical. The JiTT book recommends that the WarmUp questions share the following characteristics:

- They are motivated by a clear set of learning objectives.
- They introduce the students to the technical terms.
- They connect to students' personal real-world experience.
- They confront common naïve misconceptions.
- They are extendible.

The WarmUp assignments typically include three parts: an essay question, an estimation question, and a multiple-choice question. An example is shown in Figure 7.5. Note the interesting fact that some problems are stated ambiguously. Often we try to write questions in which all the assumptions are absolutely clear. Here, for example, it is left unstated whether the carousel is a large mechanical object in a theme park that is driven by a motor or a small unpowered rotating disk in a children's playground. Furthermore, even when you have envisioned the situation, the first part of the essay question has no unique answer. It depends on how you do it. This offers a good opportunity for starting a discussion.

The second part is a true estimation question as discussed in chapter 4. Not enough information is given (How fast are the planes going when they take to the air?), and information from personal experience must be provided (How long does it take the Earth to make one rotation?). This is also challenging since the intermediate variable required (the speed of the Earth's rotation) has to be connected to the personal data by a calculation.

Essay: Suppose you are standing on the edge of a spinning carousel. You step off, at right angles to the edge. Does this have an effect on the rotational speed of the carousel?

Now consider it the other way. You are standing on the ground next to a spinning carousel and you step onto the platform. Does this have an effect on the rotational speed of the carousel? How is this case different from the previous case?

Estimation: The mass of the Earth is about 6×10^{24} kg, and its radius is about 6×10^{6} m. Suppose you built a runway along the equator and you lined up a million 10,000 lb airplanes and had them all take off simultaneously. Estimate the effect that would have on the rotational speed of the Earth.

Multiple Choice: An athlete spinning freely in midair cannot change his

(a) angular momentum.
(b) moment of inertia.
(c) rotational kinetic energy.
(d) All of the above conclusions are valid.

Figure 7.5 A JiTT WarmUp assignment. These are distributed on the web and answered by students before they are discussed in class.

> Hold a basketball in one hand, chest high. Hold a baseball in the other hand about two inches above the basketball. Drop them simultaneously onto a hard floor. The basketball will rebound and collide with the baseball above it. How fast will the baseball rebound? Assume that the basketball is three to four times heavier than the baseball.
>
> The result will surprise you. Don't do this in the house!

Figure 7.6 A JiTT puzzle.

The multiple-choice question is not at all straightforward, though I would have offered it as a multiple-choice multiple-response question (see chapter 4), allowing the students to pick as many of the answers as they desired. A natural error here is to choose both (a) and (c), since both conservation of angular momentum and energy have been discussed. This choice is not available in the form presented. The discussion of this WarmUp cluster can be tied in class to the classic demonstration of the student with dumb-bells on the rotating chair.

A typical puzzle is given in Figure 7.6. Novak and colleagues report that most students attempting this problem get bogged down in the algebra. They then spend a full hour discussing this problem, using it as an opportunity to thoroughly review everything that had been covered to that point and to discuss and build problem-solving skills.

This example nicely illustrates the difference between JiTT questions and traditional homework problems. The goal of a JiTT question is not to evaluate students' problem-solving skills. In that case, you would hope that you had presented a question that most students can answer. In constructing JiTT questions, you want a question that is not so difficult that most students are unwilling to spend any time thinking about it, but that is hard enough that many students will not be able to complete it successfully. The primary goal of the questions is an engaged and effective lecture discussion.

The JiTT group also includes in their approach web homework of a more standard type and problems based on simulations. The book contains a brief introduction to creating simulations in the Physlet environment[7] and a set of problems and questions that can be assigned in conjunction with existing simulations.

The JiTT approach can be used in a variety of lecture-based classes and can readily be combined with other techniques in recitation and laboratory.

[7] *Physlets* is a set of programming tools using Java and JavaScript that allows the creation of simple simulations that can be delivered on the web [Christian 2001].

CHAPTER 8

Recitation and Laboratory-Based Methods

The most serious criticism which can be urged
against modern laboratory work in Physics is that
it often degenerates into a servile following of directions,
and thus loses all save a purely manipulative value.
Important as is dexterity in the handling and adjustment of apparatus,
it can not be too strongly emphasized
that it is grasp of principles, not skill in manipulation
which should be the primary object of General Physics courses.
Robert A. Millikan [Millikan 1903]

The recitation and the laboratory are two elements of the traditional structure that seem ready made for active engagement. The architectural environment can be arranged to be conducive to group work, focus on the task, and interaction. (See Figure 6.3.) Unfortunately, not much is usually done with the cognitive environment to take advantage of this opportunity. Recitations are set up with the room's movable chairs lined up as if in a large lecture hall (see Figure 10.3)—and the recitation leader does 95% of the talking. Students in laboratories may sit at tables in two groups of two, but if the lab is set up in "cookbook" style so that students can get through it quickly and without much thought, there may be little conversation and almost no effort at sense-making.

In this chapter I discuss five environments, three for recitation and two for lab.

- The traditional recitation
- *Tutorials*—Materials developed by the University of Washington Physics Education Group (UWPEG) to replace recitations by guided group concept building
- *ABP Tutorials*—Materials in the frame of those developed by the UWPEG but making use of computer-assisted data acquisition and analysis (CADAA) and video technology

- *Cooperative Problem Solving* (CPS)—An environment developed at the University of Minnesota to provide guidance for students to learn complex problem-solving skills in a group-learning environment[1]
- The traditional lab
- *RealTime Physics*—A concept-building laboratory making extensive use of CADAA

Both sets of Tutorial materials and RealTime Physics are part of the Physics Suite. The CPS materials match well with and are easily integrated with other Suite elements.

THE TRADITIONAL RECITATION

> **Environment:** Recitation.
> **Staff:** One instructor or assistant per class for a class of 20 to 30 students.
> **Population:** Introductory physics students.
> **Computers:** None.
> **Other Equipment:** None.
> **Time Investment:** Low.

The traditional recitation has an instructor (at the large research universities this is often a graduate student) leading a one-hour class for 20 to 30 students. These sections are often tied to the homework: students ask questions about the assigned problems, and the teaching assistant (TA) models the solution on the blackboard. If the students don't ask questions about any particular problems, the TA might choose problems of his or her own and model those. A brief quiz (10 to 15 minutes) may be given to make sure students attend. At Maryland, this regime has been standard practice for decades. Sometimes, due to time pressures and a limited number of TAs, homework grading is dropped, and the quiz is one of the homework problems chosen at random—to guarantee that the students have to do them all, even though they aren't collected. The recitation becomes a noninteractive lecture in which the students are almost entirely passive.

When I first taught the calculus-based physics class about a dozen years ago, I asked Sagredo for his advice. "Problem solving is really important," he responded, "so be sure your TAs give a quiz to bring the students into the recitation on a regular basis." I was intrigued by the assumption implicit in this statement: that the activity was important for student learning but that without a compulsion, students would not recognize this fact.

I decided to test this for myself. I told my students that in recitation, the TAs would be going over problems of the type that would appear on the exams. They would not be required to come, but it would help them do better on the exams. The result was the disaster that Sagredo had predicted. Attendance at the recitations dropped precipitously. When, about

[1]The CPS project has also developed a laboratory curriculum that articulates with the problem-solving recitations, putting each laboratory exercise into the context of a problem. These laboratories are not discussed in detail here. For more information, see [Heller 1996] and the group's website at http://www.physics.umn.edu/groups/physed.

halfway into the semester, I asked one of my TAs how his attendance was, he remarked: "It was great last week. I actually had eight students show up [out of a class of 30]." Now and then I stood outside one of these recitation rooms to listen to what was going on. It seemed that there were two or three students in the group who were on top of things, had tried to do the homework and had real questions, and were following closely. Then, there were another three or four students who didn't say a word but were writing down everything that was said. My assumption is that they were "pattern matchers"—students who did not assume that it was necessary to understand or make sense of the physics and felt they could get by with memorizing a bunch of problems and then replaying them on the exam. This impression was reinforced by my interaction with these students during office hours.

The next time I taught the class I decided that since the students didn't see recitations as valuable to their learning, perhaps they were right. I eliminated the recitation in favor of a group-learning concept-building activity, *Tutorials* [Tutorials 1998].[2] I told them that in Tutorial we would be working through basic concepts. They would not be required to come, but it would help them do better on the exams. Interestingly enough, despite the similarity of the instructions, the attendance results were dramatically different. The TAs reported almost full classes (80% to 95%) at every session.[3] I don't fully understand the psychology behind this, but my first guess is that the social character of the Tutorial classes changed the way they thought about the class. Since they were interacting with their peers, the activity was no longer individual and they had some responsibility for being there to interact. Put another way, Tutorials are like laboratories and one did not cut a lab if one could help it, in part because it caused a serious problem for your lab partner. The traditional recitations are more like lectures, and nobody really cared if you missed lecture.

A more interactive approach to the traditional recitation

Even if you don't want to (or have the resources to) implement a research-based recitation replacement such as Tutorials, in a small class of 20 to 30 you can use many techniques to increase the students' engagement with the material. The small-class environment provides lots of opportunities for this engagement. Some methods include:

- *Ask authentic questions*—Questions that are relevant to what the students are learning and that you expect them to answer are much more engaging than rhetorical questions or questions that interest only one student.
- *Lead a discussion*—Don't answer student questions yourself, but see if you can get a discussion going to answer the questions. Help them along now and then if needed.
- *Have them work on problems together*—Problems that are not assigned for homework but that rely on an understanding of fundamental concepts can be effective and engaging. (Problems that only rely on straightforward algebraic manipulations are not.) Having one student from each group put their solution on the board and then having a class

[2] I use the word "Tutorials" with a capital T to distinguish the specific University of Washington style of lessons from a more traditional "tutorial" in which a student is tutored—perhaps led through a lesson step by step. "Capital T" Tutorials are a more complex activity.

[3] Early morning (8 A.M.) sessions are sometimes an exception.

discussion can be very valuable. Some of the context-based reasoning problems discussed in chapter 4 can be effective here.

- *Do fewer problems and go into them more deeply*—If you do many problems quickly, it encourages the students' view that they need to pattern match rather than understand. Going through a problem of medium difficulty with enough discussion that student confusions are revealed may take 20 to 30 minutes.

These approaches sound easier than they are. Each one succeeds better the more you understand about where the students actually are—what knowledge they bring to the class, both correct and incorrect, and what resources they have to build a correct knowledge structure. The critical element is communication.

Redish's ninth teaching commandment: Listen to your students whenever possible. Give them the opportunity to explain what they think and pay close attention to what they say.

Helping your teaching assistants give better recitations

For instructors in charge of a group of TAs, I have some additional words of advice.

- *Make sure that your TAs understand the physics*—Faculty have a tendency to assume that graduate students are well versed in introductory physics. But remember: they may be novice TAs and have last studied introductory physics four or five years ago. Little of what they have done since then (Lagrangians, quantum physics, Jackson problems) will help them with the often subtle conceptual issues in an introductory class.
- *Make sure that you and your TAs are on the same page*—If you are trying to stress conceptual issues and promote understanding, make sure that your TAs know what you are trying to do and understand it. If you want them to use a particular method to solve a class of problems, be sure the TAs know that you are pushing it.
- *Worry about grading and administrative details*—One of the easiest ways to get in trouble with your students is to have different TAs grading them in different ways. A TA who grades homework casually, giving points for effort, can produce a pattern of scores much higher than one who slashes points for trivial math errors. This can cause difficulty in assigning grades fairly and can lead to significant student anger and resentment.

These guidelines are based on my experience. One should be able to create an effective and engaging learning environment in a class of 20 to 30 students, even without adopting a special curriculum. However, careful research has yet to be done to see what elements are critical in producing effective learning in this situation.

Studies in other environments suggest that even when the instructor is sensitive to research-determined difficulties that students have with the material, research-based instructional materials may make a big difference in the effectiveness of the instruction.[4] In the next two sections I describe three elements that can transform recitations into more effective learning environments.

[4]See [Cummings 1999].

TUTORIALS IN INTRODUCTORY PHYSICS

Environment: Recitation.

Staff: One trained facilitator per 15 students.

Population: Introductory physics students. (All are appropriate for calculus-based physics classes; some are also appropriate for algebra-based classes.)

Computers: Very limited use.

Other Equipment: Butcher paper or whiteboards and markers for each group of three to four students. Occasional small bits of laboratory equipment for each group (e.g., batteries, wires, and bulbs).

Time Investment: Moderate to substantial (one to two hour weekly training of staff required).

Available Materials: A manual of tutorial worksheets and homework.

Perhaps the most carefully researched curriculum innovation for introductory calculus-based physics is *Tutorials in Introductory Physics*, developed by Lillian C. McDermott, Peter Shaffer, and the University of Washington Physics Education Group (UWPEG). Numerous Ph.D. dissertations by students in this group have extensively investigated student difficulties with particular topics in calculus-based physics and have designed group-learning lessons using the research-and-redevelopment wheel. (See Figure 6.1.) References to this extensive body of work can be found in the Resource Letter included in the Appendix of this volume. The published materials cover a wide range of topics from kinematics to physical optics [Tutorials 1998]. Additional materials are continually being developed and refined.

In *Tutorials*, the traditional recitation is replaced by a group-learning activity with carefully designed research-based worksheets. These worksheets emphasize concept building and qualitative reasoning. They make use of cognitive conflict and bridging, and use trained facilitators to assist in helping students resolve their own confusions. The method can be implemented to help improve student understanding of fundamental physics concepts in a cost-effective manner within the traditional lecture structure [Shaffer 1992] [McDermott 1994].

Students in Tutorials work in groups of three to four with a wandering facilitator for every 12 to 15 students. These facilitators check the students' progress and ask leading questions in a semi-Socratic dialog[5] to help them work through difficulties in their own thinking. (See Figure 8.5.) The structure of the classroom (Figure 6.3) reflects the different focus in behavior expected in a Tutorial as compared to a lecture. In this, and in other inquiry-based classes, the student's focus is on the work (on the table) and on the interaction with the other students in their group.

[5]See Bob Morse's lovely little article "The Classic Method of Mrs. Socrates" to learn more about the difference between a "Socratic" and a "semi-Socratic" dialog [Morse 1994].

The structure of Tutorials

Tutorials have the following components:

1. A 10-minute ungraded "pre-test" is given once a week (typically in lecture). This test asks qualitative conceptual questions about the subject to be covered in Tutorial the following week and gets the students thinking about some (usually counterintuitive) issues.
2. The teaching assistants and faculty involved participate in a one- to two-hour weekly training session. In this session, the TAs do the pre-test themselves and go over the students' pre-tests (but don't grade them). They discuss where the students have trouble and do the Tutorial in student mode, nailing down the (often subtle) physics ideas covered.
3. Students attend a one-hour (50-minute) session. Students work in groups of three or four and answer questions on a worksheet that walks them through building qualitative reasoning on a fundamental concept.
4. Students have a brief qualitative homework assignment in which they explain their reasoning. This helps them bring back and extend the ideas covered in the Tutorial. It is part of their weekly homework, which in most cases also includes problems assigned from the text.
5. A question emphasizing material from Tutorials is asked on each examination. (See Redish's sixth teaching commandment!)

Tutorials often focus on important but subtle points

At the University of Washington, tutorial worksheets are developed over a period of many years using the research-redevelopment cycle. The UWPEG has a highly favorable situation for curriculum development at the University of Washington—a large research group of graduate students and postdocs, continued support for research and development over many years, and an educational environment in which every term of the calculus-based physics course is taught using Tutorials four times a year. As a result, the UW Tutorials are highly refined and very carefully thought out and tested. Although one might think one sees some obvious "fixes," I recommend that they not be changed lightly.

When we first introduced Tutorials at the University of Maryland, Sagredo was lecturing one of the sections in which we were testing tutorials. He suggested that the vector acceleration activity shown in Figure 8.1 be replaced by motion on a circle. "After all," he commented, "circular motion is much simpler than elliptical motion, so they should understand it better."

Sagredo's comment misses the point of the activity. Reif and Allen have demonstrated [Reif 1992] that students often don't internalize the concept of vector acceleration well at all. They try to memorize formulas that will allow them to solve problems without struggling to make sense of the fundamental concepts. In the case of acceleration, the students needed to learn to think of vector acceleration through a process—looking at velocity vectors at nearby times and seeing how they changed. The activity in the UW tutorial is carefully designed to

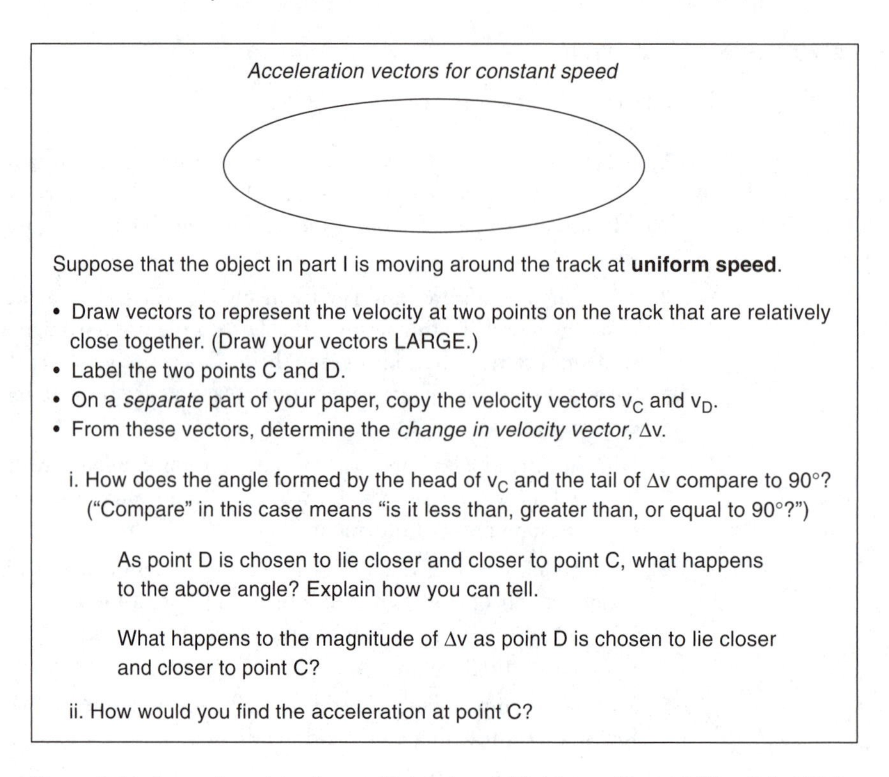

Acceleration vectors for constant speed

Suppose that the object in part I is moving around the track at **uniform speed**.

- Draw vectors to represent the velocity at two points on the track that are relatively close together. (Draw your vectors LARGE.)
- Label the two points C and D.
- On a *separate* part of your paper, copy the velocity vectors v_C and v_D.
- From these vectors, determine the *change in velocity vector*, Δv.

i. How does the angle formed by the head of v_C and the tail of Δv compare to 90°? ("Compare" in this case means "is it less than, greater than, or equal to 90°?")

As point D is chosen to lie closer and closer to point C, what happens to the above angle? Explain how you can tell.

What happens to the magnitude of Δv as point D is chosen to lie closer and closer to point C?

ii. How would you find the acceleration at point C?

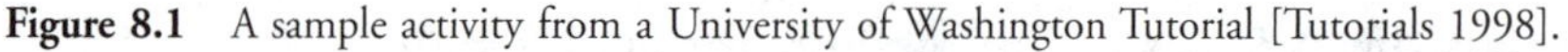

Figure 8.1 A sample activity from a University of Washington Tutorial [Tutorials 1998].

be sufficiently general (*not* a circle, *not* an ellipse) so that the students can't pattern match to something in the book, but sufficiently specific (moving at a constant speed but with changing direction, later with changing speed) to force them to focus on the process of constructing the acceleration. Following Sagredo's well-meaning advice would have completely undermined the carefully designed learning activity.

Should you post solutions to Tutorial pre-tests and homework?

The UW Tutorials often rely on the cognitive conflict method discussed in chapter 2. In this approach, situations are presented that cue common student difficulties revealed by research. The facilitators then help those students who show the predicted difficulties work through their ideas themselves. McDermott refers to this process as *elicit/confront/resolve* [McDermott 1991]. The pre-tests often raise questions that appear to be straightforward and many (if not most) of the students miss. Note that the pre-tests should *not* be gone over in class, nor should the results be posted. The point of the pre-tests is to get the students thinking about the issues. They then confront these issues for themselves during the tutorial session. Giving them the answers short-circuits the learning activity.

Sagredo was worried about this and after a few weeks of Tutorials, asked his students in lecture whether they wouldn't like to have the answers to the pre-tests posted. The result was

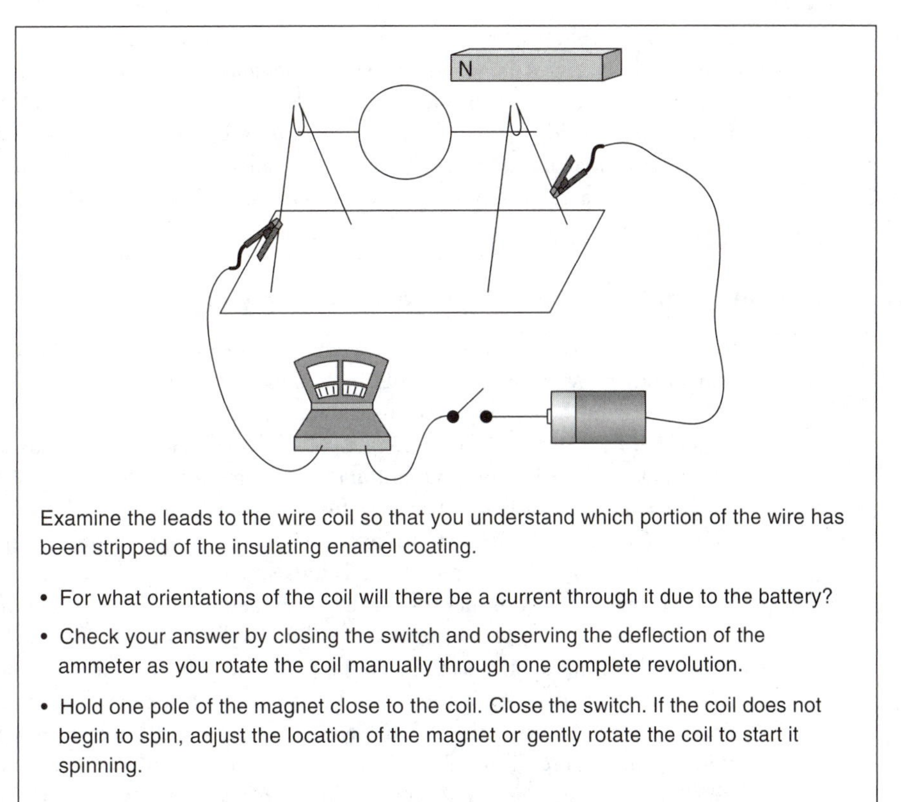

Figure 8.2 An activity from a UW Tutorial involving simple equipment [Tutorials 1998].

lukewarm. One of the students spoke up and said, "Well, we just go over the answers in Tutorial the next week so we get it there."

The answers to the Tutorial homework may be a different story. The Tutorial homework is supposed to be a reasonably straightforward extrapolation of what was done in Tutorials in order to provide reinforcement at a later time. In some of my classes, this did not appear to be a problem, especially when most groups had a chance to finish the Tutorials. In classes where the Tutorials often were not finished (e.g., in my algebra-based classes), students found the Tutorial homework more difficult. When I discovered that some of my graduate assistants had gotten the answers wrong on the Tutorial homework, I decided to provide solutions.

What does it take to implement Tutorials?

The UW Tutorials focus on fundamental concepts and low implementation costs. Although they occasionally call for a few items of easily obtained equipment (batteries and bulbs, magnets, compasses), most are done with paper and pencil as in the example shown in Figure 8.1. Some of the activities, however, involve well-designed and exciting mini-labs. (See, for example Figure 8.2.) The largest investment required is that someone has to become an expert in the method and provide training for the facilitators and someone has to manage the operation.[6]

The tutorial and homework sheets are available for purchase by the students [Tutorials 1998]. Pre-tests, sample examination problems, and equipment lists are available with the instructor's guide.

Tutorials produce substantially improved learning gains

Tutorials have been extensively researched and tested by the University of Washington group and by others. Many of the UWPEG's publications over the past decade have been research associated with the development of specific Tutorials.

We carried out a test of a secondary implementation of Tutorials at the University of Maryland using the FCI pre and post in the first semester of engineering physics [Redish 1997]. To see the range of variation that arose from different lecturers, Saul, Steinberg, and I gave the FCI to 16 different lecture sections involving 14 different professors. Seven of the sections used traditional recitations, and nine used Tutorials. The classes chosen to use Tutorials were selected at random. Two professors taught twice, once with Tutorials and once without.

The fraction of the possible gain,[7] g, attained averaged 0.20 in the lecture classes and 0.34 in the Tutorial classes. Each of the professors who taught with and without Tutorials each had better scores with Tutorials by 0.15. (One of these professors taught with Tutorials first, one with recitation first.) A histogram of the results is shown in Figure 8.3. Every lecture section that used Tutorials achieved a higher value of g than every section that used recitations. (A later class with an award-winning professor achieved a gain of 0.34 without Tutorials, higher than our lowest gain with Tutorials but lower than most of the Tutorial-based classes.)

Changing recitations to Tutorials doesn't hurt problem solving

In their dissertations at Maryland, Jeff Saul and Mel Sabella studied problem solving in our tutorial/recitation comparison. In most cases, there was little or no difference observed between the two groups on traditional exam problems. On a few problems, the tutorial students did dramatically better than those in recitations. The interesting cases are those where the tutorial did not specifically cover the kind of example in the problem but appeared to help students build a functional mental model.

An example of such a problem is shown in Figure 8.4, and the results at Maryland are shown in Table 8.1 [Ambrose 1998]. The problem is trickier than it looks. Students who memorize equations tend to memorize them in the order given in the book, and the one for the position of the bright fringes is always given first. The problem, however, asks for the position of the *dark* fringe. A large fraction of the students in the recitation class simply pulled out the bright-fringe formula and got an answer off by a factor of 2.

[6] At Maryland, we found that the extra costs needed to run Tutorials for 600 students corresponded to about one-half of a teaching assistant—the person needed to organize and manage the operation.

[7] See the discussion of g in chapter 5 for a definition.

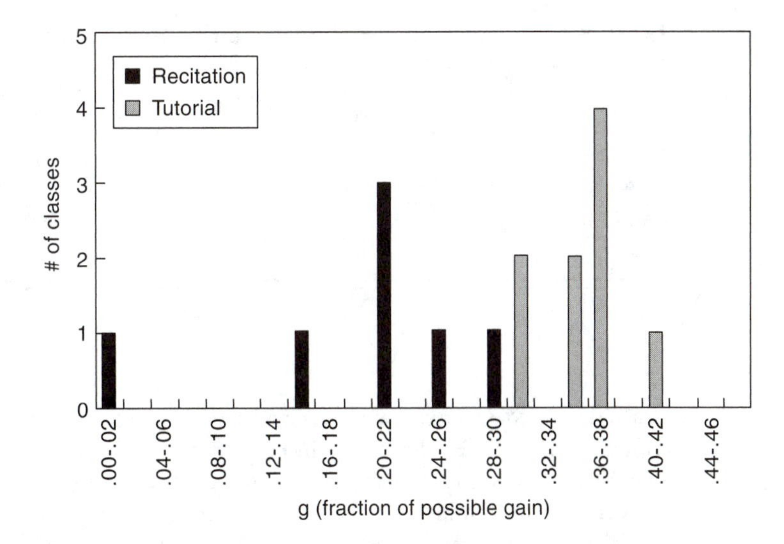

Figure 8.3 Fraction of the possible gain attained by engineering physics students at the University of Maryland in classes taught with traditional recitations (dark) and tutorials (light).

A satisfyingly large fraction of the students in the Tutorial class actually reasoned their way to the answer using a path-length argument, showing that they could call on an underlying mental model to construct a correct result. I was particularly impressed with this result since the Tutorials we used do not explicitly consider this problem but focus on building the concept of path length and its role in interference.

Students need to get used to Tutorials

In introducing Tutorials to a class, you should be aware of possible attitudinal difficulties. As discussed in chapter 3, students bring to their physics classes expectations about the type of knowledge they will be learning in class and what they have to do to get it. Engineering students (especially those who have taken AP physics in high school) may have a strong expectation that what they are supposed to learn in a physics class are equations and how to produce numbers. The idea of "concepts" and even the idea of "making sense" of anything in physics may be foreign to them. These students can at first be quite hostile to the idea behind Tutorials. Some of the better students think Tutorials are trivial (despite making numerous errors in their predictions). Others may be accustomed to operating in a competitive

> Light with $\lambda = 500$ nm is incident on two narrow slits separated by $d = 30\ \mu$m. An interference pattern is observed on a screen a distance L away from the slits. The first dark fringe is found to be 1.5 cm from the central maximum. Find L.

Figure 8.4 A problem on which Tutorial students performed significantly better than recitation students [Ambrose 1998].

TABLE 8.1 Results on Problem Given to Recitation and Tutorial Classes

	Example	Recitation ($N = 165$)	Tutorial ($N = 117$)
$L = 1.8$ m (correct)	$\Delta D = d \sin\theta = \lambda/2$ $\sin\theta = y/L$	16%	60%
$L = 0.9$ m	$y = m\lambda L/d$	40%	9%
Other	$L = 5.0 \times 10^{-7}$ m	44%	31%

rather than a cooperative framework and may not like "having to explain their answers to dummies." (I've gotten this comment even after a session when one of the "dummies" asked a probing question that helped that overconfident self-categorized "top student" correct a serious error in his thinking.) Once both the faculty and the student body come to accept Tutorials as a normal part of the class, Tutorials tend to be rated as one of the most valuable parts of the class.

Given that conceptual learning and qualitative reasoning may be new to many of the students in an introductory physics class, the introduction of Tutorials needs to be done carefully. I have had the most success when I have integrated the Tutorial approach fully into my lectures and tied qualitative reasoning to my problem-solving examples. Exams that contain a "Tutorial question" are a minimum necessity. Exams in which every question blends Tutorial ideas with problem solving are even more effective in helping students understand the value of concepts and qualitative thinking.

ABP TUTORIALS

Environment: Recitation.

Staff: One trained facilitator per 15 students.

Population: Introductory calculus-based physics students. (Many of the tutorials are also appropriate for algebra-based classes.)

Computers: One for every three to four students.

Other Equipment: Butcher paper or whiteboards and markers for each group of three to four students. Occasional small bits of laboratory equipment for each group (e.g., batteries, wires, and bulbs). Computer-assisted data acquisition device; various programs and simulations including *Videopoint* and *EM Field*.

Time Investment: Moderate to substantial (one to two hours weekly training of staff required)

Available Materials: A set of tutorial worksheets, pre-tests, and homework. [ABP-Tutorials] These tutorials and instructions for their use are available on the web at http://www.physics.umd.edu/perg/.

Although the UWPEG Tutorials cover a wide range of topics, they strongly focus on the issue of qualitative reasoning and concept building. In addition, the UWPEG made the choice to make Tutorials as easy to implement as possible, so they rely on very little (and very inexpensive) equipment. In addition, the UWPEG Tutorials are designed so that they can be reasonably successful in helping students build fundamental concepts even if concept building is not significantly supported elsewhere in the course (lecture, laboratory, homework problems). One difficulty with such a situation is that students tend to develop independent schemas for qualitative and quantitative problem solving and to only occasionally (as discussed in the section on Tutorials) cross qualitative ideas over to help them in solving quantitative problems [Kanim 1999] [Sabella 1999].

ABP Tutorials are mathematically and technologically oriented

The University of Maryland Physics Education Research Group (UMdPERG), as part of the Activity-Based Physics (ABP) project, developed a supplementary set of Tutorials that are based on a different set of assumptions [Steinberg 1997]:

1. We assume that conceptual learning is being integrated throughout the course—in lecture, homework, and laboratories—so that Tutorials are not the sole source of conceptual development.
2. We assume that quantitative problem solving is a significant goal of the class.
3. We assume that reasonable computer tools are available for use in Tutorials.

Given these assumptions, the structure of Tutorial lessons can be changed somewhat. They can focus more on relating conceptual and mathematical representations and on building qualitative to quantitative links. In this set of lessons, computers are used for taking data, displaying videos, and displaying simulations. For example, in Figure 8.5, a facilitator

Figure 8.5 Interactive computer-based tutorial on Newton's law. Students are interacting with the facilitator (standing), who asks Socratic-dialog questions to focus their inquiries.

(standing) is shown talking to a group of students who are using a fan cart on a Pasco track with a sonic ranger detector to study Newton's second law.

While some of the lessons are new, others have been adapted to the Tutorial framework from *RealTime Physics* laboratories and from *Workshop Physics* activities. These include

- Discovering Newton's third law using two force probes on Pasco carts
- Exploring the concept of velocity by walking in front of a sonic ranger
- Exploring oscillatory behavior by combining a mass on a spring hanging from a force probe with a sonic ranger

New lessons include such topics as

- Tying the concepts of electric field and electrostatic potential to their mathematical representations using the software *EM Field* [Trowbridge 1995]
- Building an understanding of the functional representation of wave propagation using video clips of pulses on springs and the video analysis program *Videopoint*™ [Luetzelschwab 1997]
- Building an understanding of the oscillatory character of sound waves and the meaning of wavelength using video clips of a candle flame oscillating in front of a loudspeaker and the video analysis program *Videopoint*™ (See Figure 8.6.)

Concept learning can be tied to the use of math

The example of sound waves gives an interesting example of how concept learning can be tied to mathematical concepts by using media effectively.

Figure 8.6 A frame from a video used in an ABP Tutorial. A low-frequency sound wave (10 Hz) emitted by the speaker causes the candle flame to oscillate back and forth. Students use *Videopoint*™ to measure the frequency of the oscillation.

Students studying the topic of sound in the traditional way often construct a picture of a sound wave that treats the peaks of the wave (or pulse) as if it were a condensed object pushed into the medium rather than as a displacement of the medium [Wittmann 2000]. Wittmann refers to these two mental models as *the particle pulse model* (PP) and the *community consensus model* (CC). One clue that a student is using a PP model is that the student assumes that each pulse of sound that passes a floating dust particle "hits" it and pushes it along.

Alex: *[The dust particle] would move away from the speaker, pushed by the wave, pushed by the sound wave. I mean, sound waves spread through the air, which means the air is actually moving, so the dust particle should be moving with that air which is spreading away from the speaker.*

Interviewer: *Okay, so the air moves away—*

A: *It should carry the dust particle with it.*

I: *How does [the air] move to carry the dust particle with it?*

A: *Should push it, I mean, how else is it going to move it? [sketches a typical sine curve] If you look at it, if the particle is here, and this first compression part of the wave hits it, it should move it through, and carry [the dust particle] with it.*

. . .

I: *So each compression wave has the effect of kicking the particle forward?*

A: *Yeah.*[8]

In his thesis research, Wittmann found that most students in engineering physics used the PP model most of the time but used the CC model in some circumstances. One ABP Tutorial developed to deal with this issue uses a video of a flame in front of a loudspeaker.

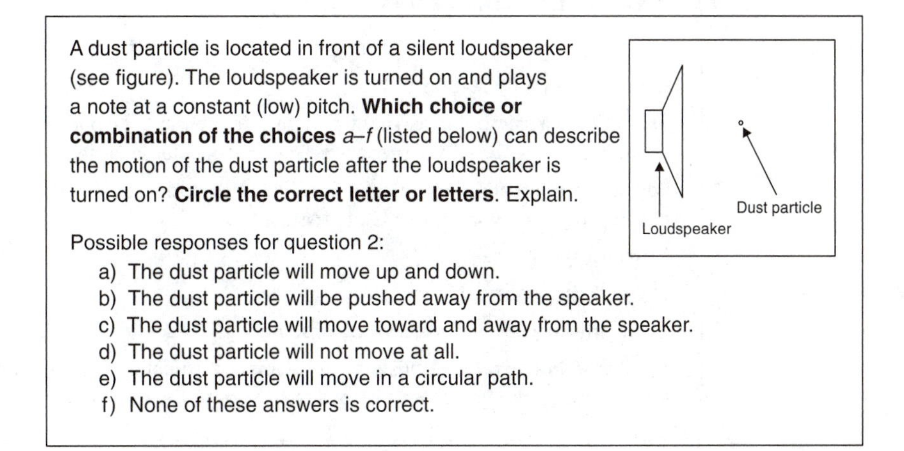

A dust particle is located in front of a silent loudspeaker (see figure). The loudspeaker is turned on and plays a note at a constant (low) pitch. **Which choice or combination of the choices** *a–f* (listed below) can describe the motion of the dust particle after the loudspeaker is turned on? **Circle the correct letter or letters**. Explain.

Possible responses for question 2:

a) The dust particle will move up and down.
b) The dust particle will be pushed away from the speaker.
c) The dust particle will move toward and away from the speaker.
d) The dust particle will not move at all.
e) The dust particle will move in a circular path.
f) None of these answers is correct.

Figure 8.7 MCMR problem used to probe mental models students use in envisioning sound.

[8]Dialog quoted from [Wittmann 2001a].

TABLE 8.2 Student Performance on Sound Wave Questions Before, After Traditional Lecture, and After Additional Modified Tutorial Instruction

Time during Semester MM used	Before All Instruction (%)	Post-Lecture (%)	Post-Lecture, Post-Tutorial (%)
CC (longitudinal oscillation)	9	26	45
Other oscillation	23	22	18
PP (pushed away linearly or sinusoidally)	50	39	11
Other	7	12	6
Blank	11	2	21

(See Figure 8.6.) Using cognitive conflict methods, the lesson has students predict the flame's behavior when the speaker is turned on. Students then track how the flame moves back and forth and create a graph of the tip's oscillatory motion. Then they consider what will happen when a sound wave passes a chain of separated flames in order to build an understanding of the idea of relative phase and wavelength.

An exam question used to test the students' responses to this lesson is shown in Figure 8.7, and the results for traditional and Tutorial instruction are presented in Table 8.2. Data are matched ($N = 137$ students). The large number of blank responses in the post-all instruction category is due to the number of students who did not complete the pre-test on which the question was asked. The results show that Tutorials are a significant improvement over traditional instruction.[9]

COOPERATIVE PROBLEM SOLVING (CPS)

Environment: Recitation.

Staff: One instructor or assistant per class trained in the approach for any number of students (N = 20–30); a second facilitator is helpful in larger classes.

Population: Introductory calculus-based physics students. (Many of the problems are also appropriate for algebra-based classes.)

Computers: None.

Other Equipment: None.

Time Investment: Moderate to substantial.

Available Materials: Manuals of problems for students and an instructor's guide. Available from the group website at http://www.physics.umn.edu/groups/physed.

[9] The development environment at Maryland does not match the ideal one at Washington, so these Tutorials have only been through two to four cycles of development compared to the eight to ten typically carried out at Washington. As a result, they are not as refined—and not as effective.

Over the past decade or so, Pat and Ken Heller at the University of Minnesota and their collaborators have developed a group-learning problem-solving environment in which students work together in recitation on problems they have not previously seen [Heller 1992]. Their work is based on the generalized studies of the Johnsons and their collaborators on the effectiveness of group learning [Johnson 1993].

Cooperative Problem Solving relies on context-rich problems

The problems the Minnesota group have developed are *context rich*; that is, they involve realistic situations, may contain incomplete data, and may require the students to pose a part of the problem themselves. (See Figure 8.8.) The problems are intended to be too difficult for any individual student to solve but not too hard for a group of students of mixed ability to solve in about 15 to 20 minutes when working together. Groups are formed to include students of varying ability, and students may be assigned specific (and rotating) roles to play in each group.

The Minnesota group's context-rich problems have a number of general characteristics that encourage appropriate thinking. These characteristics may be difficult for a novice problem-solver to handle individually, and they facilitate discussion. They include the following:

1. It is difficult to use a formula to plug in numbers to get an answer.
2. It is difficult to find a matching solution pattern to get an answer.
3. It is difficult to solve the problem without first analyzing the problem situation.
4. It is difficult to understand what is going on in this problem without drawing a picture and designating the important quantities on that picture.
5. Physics words such as "inclined plane," "starting from rest," or "projectile motion" are avoided as much as possible.
6. Logical analysis using fundamental concepts is reinforced.

> A friend of yours, a guitarist, knows you are taking physics this semester and asks for assistance in solving a problem. Your friend explains that he keeps breaking repeatedly the low E string (640 Hz) on his Gibson "Les Paul" when he tunes up before a gig. The cost of buying new string is getting out of hand, so your friend is desperate to resolve his dilemma. Your friend tells you that the E string he is now using is made of copper and has a diameter of 0.063 inches. You do some quick calculations and, given the neck of your friend's guitar, estimate the the wave speed on the E string is 1900 ft/s. While reading about stringed instruments in the library, you discover that most musical instrument strings will break if they are subjected to a strain greater than about 2%. How do you suggest your friend solve his problem?

Figure 8.8 Sample of a context-rich problem from the Minnesota CPS method [Heller 1999].

The problems are written so as to require careful thinking.

- The problem is a short story in which the major character is the student. That is, each problem statement uses the personal pronoun "you."
- The problem statement includes a plausible motivation or reason for "you" to calculate something.
- The objects in the problems are real (or can be imagined)—the idealization process occurs explicitly.
- No pictures or diagrams are given with the problems. Students must visualize the situation by using their own experiences.
- The problem solution requires more than one step of logical and mathematical reasoning. There is no single equation that solves the problem.

These types of problems change the frame: students cannot simply "plug-and-chug" or pattern match. They have to think about the physics and make decisions as to what is relevant. This strongly encourages the group to try to make sense of the problem rather than simply come up with the answer.

The Minnesota group does not simply drop these harder problems on their students. They develop an explicit problem-solving strategy, and they help the students apply it when they get stuck. The broad outlines are illustrated in Figure 8.9.[10]

Although it's hard to create such problems, when you get one, the effect of the group interaction can be quite dramatic. But the whole idea of solving problems in a group may be difficult—for the TAs as much as for the students. One year I prepared some of these problems and handed one out to my TAs on transparencies each week, in order to encourage them to begin some group work instead of lecturing to the students. One TA, having had no experience with group work herself, decided not to follow my instructions. Instead of assigning the problem as group work, she presented it as a quiz at the beginning of the class. When after 10 minutes most of the students said they had no idea how to begin, she let them use their class notes. When after an additional five minutes they still were not making progress, she let them open their texts. After 20 minutes, she collected and graded the quiz. The results were awful, the average being about 20%. The students in her class complained that "the quizzes were too hard." When I questioned her about what she had done, she replied that "If they work together, you don't know who's responsible for the work." She had mistaken an activity which I had intended to serve a teaching purpose for one meant to serve as an assessment. In other sections, many groups solved the problem successfully.

Group interactions play a critical role

Sagredo had some sympathy for my TA. "If they work together, you'll just get to see the work of the best student. The weaker students will just go along for the ride," he complained. The

[10]This strategy is an elaboration of the strategy found in Polya's famous little book, *How to Solve It* [Polya 1945]. The Minnesota group found that using Polya's strategy directly was too difficult for the algebra-based class and that some intermediate elaborations were required [Heller 1992]. Polya's strategy is: (1) understanding the problem, (2) devising a plan, (3) carrying out the plan, and (4) looking back.

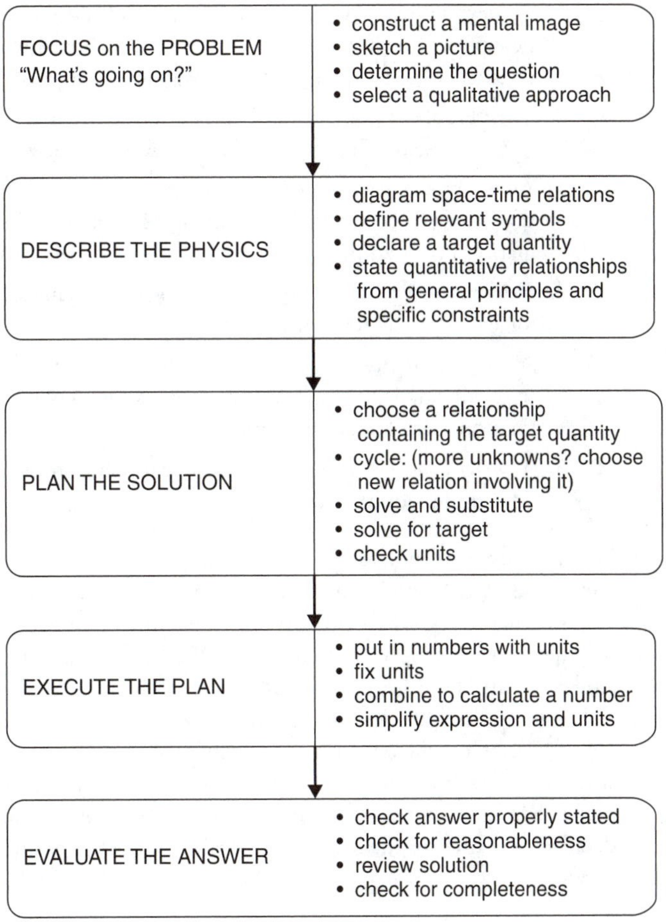

FOCUS on the PROBLEM "What's going on?"	• construct a mental image • sketch a picture • determine the question • select a qualitative approach
DESCRIBE THE PHYSICS	• diagram space-time relations • define relevant symbols • declare a target quantity • state quantitative relationships from general principles and specific constraints
PLAN THE SOLUTION	• choose a relationship containing the target quantity • cycle: (more unknowns? choose new relation involving it) • solve and substitute • solve for target • check units
EXECUTE THE PLAN	• put in numbers with units • fix units • combine to calculate a number • simplify expression and units
EVALUATE THE ANSWER	• check answer properly stated • check for reasonableness • review solution • check for completeness

Figure 8.9 Structure of the problem-solving method used by the Minnesota group (simplified and condensed somewhat) [Heller 1999].

Minnesota group has shown that this is not the case, and they have developed techniques to improve group interactions.

The work of the group is better than the work of the best student in it

In order to evaluate the success of groups compared to the success of the best individual in the group, the Minnesota group compared individual and group problem-solving success [Heller 1992]. Since you can't give the same problems to the same students in different contexts and compare the results, they developed a scheme for determining problems that had approximately the same level of difficulty. They classified problem difficulty by considering six characteristics.

1. *Context:* Problems with contexts familiar to most students (through direct experience, newspapers, or TV) are less difficult than those involving unfamiliar technical contexts (such as cyclotrons or X-ray signals from pulsars).
2. *Cues:* Problems containing direct cues to particular physics (mention of force or "action and reaction") are less difficult than those for which the physics must be inferred.
3. *Match of given information:* Problems with extraneous information or information that needs to be recalled or estimated are more difficult than those where the information provided precisely matches the information needed.
4. *Explicitness:* Problems where the unknown required is specified are easier than those for which it has to be invented.
5. *Number of approaches required:* Problems that only need one set of related principles (e.g., kinematics or energy conservation) are less difficult than those requiring more than one set of such principles.
6. *Memory load:* Problems that require the solution of five equations or fewer are easier than those requiring more.

For each problem they assigned a difficulty value of 0 or 1 on each of these characteristics and found that the problem score was a good predictor of average student performance.

In order to test whether the groups were operating effectively to produce better solutions or whether they simply represented the work of the group's best student, they tested their students using both group and individual problems. Problems were matched as to level of difficulty on the scale described above. Over six exams in two terms, the groups averaged 81 ($N = 179$), while the best-in-group individual had an average of 57. These results were consistent over the different exams and classes and strongly suggest that the groups are performing better than the best individual in the group.[11] By now, the group has developed a much more detailed structure for identifying the difficulty level of a question. See the group website for details.

Techniques for improving group interactions

The Minnesota group has studied the dynamics of group interactions in CPS and has developed a number of recommendations:

- *Assign roles:* In order to combat the tendency of students to select narrow roles during group activities and therefore limit their learning, the Minnesota group assigns roles to students in the group: manager, explainer, skeptic, and recordkeeper. These roles rotate throughout the semester.
- *Choose groups of three:* Groups of two were not as effective in providing either conceptual or procedural knowledge as groups of three or four. In groups of four, one student sometimes tended to drop out—either a timid student unsure of him/herself, or a good student tired of explaining things.

[11] Students were given unlimited time to complete the final exam. Their comparative study of incomplete problems in the midsemester (time limited) and final exams shows that time considerations do not upset this conclusion [Heller 1992].

- *Assign groups to mix ability levels:* Groups with strong, medium, and weak students performed as well as groups containing strong students only. Often, the questions asked by weaker students helped strong students identify errors in their thinking. Groups of uniformly strong students also tended to overcomplicate problems.
- *Watch out for gender problems:* Groups of two males and one female tend to be dominated by the males even when the female is the best student in the group.
- *Help groups that are too quick to come to a conclusion:* This can occur when a dominant personality railroads the group or because of a desire to quickly accept the first answer given. Some groups try to quickly resolve disagreements by voting instead of facing up to their differences and resolving them.

I have noted this last problem in Tutorials as well. In both cases, facilitators can help get them back on track, encouraging them to reconsider and resolve discrepancies. The Minnesota group suggests that group testing can help with this problem.

In Jeff Saul's dissertation, he studied four different curricular innovations including cooperative problem solving [Saul 1998]. He observed both the University of Minnesota's implementation in calculus-based physics and a secondary implementation at the Ohio State University. Pre-post testing with the FCI indicated that CPS was comparable to Tutorials in producing improvements in the student's conceptual understanding of Newtonian mechanics. (See Figure 9.4.) This is interesting since CPS focuses on quantitative rather than qualitative problem solving.

Unfortunately, Saul found no significant gains on the MPEX survey.[12] So even though students appeared to improve their conceptual knowledge (and their ability to use that knowledge), their conscious awareness of the role of concepts did not seem to improve correspondingly.

THE TRADITIONAL LABORATORY

Environment: Laboratory.

Staff: One instructor or assistant per class trained in the approach for a class of 20 to 30 students.

Population: Introductory physics students.

Computers: If desired, one for every pair of students.

Other Equipment: Laboratory equipment.

Time Investment: Medium.

The laboratory is the single item in a traditional physics course where the student is expected to be actively engaged during the class period. Unfortunately, in many cases the laboratory has turned into a place to either "demonstrate the truth of something taught in lecture" or

[12]See the discussion of the MPEX in chapter 5.

to "produce a good result." The focus in both of these cases is on the content and not on what might be valuable for a student to learn from the activity. In the United States, "cookbook" laboratories—those in which highly explicit instructions are given and the student doesn't have to think—are common. They are unpopular with students and tend to produce little learning. Some interesting "guided-discovery labs" have been developed in the past few years that appear to be more effective.

Despite some interesting research on learning in laboratories in the early years of PER (e.g., [Reif 1979]) and a few recent studies (e.g., [Allie 1998] and [Sere 1993]), there has been little published research on what happens in university physics laboratories.

Goals of the laboratory

One can imagine a variety of goals for a laboratory:

- *Confirmation*—To demonstrate the correctness of theoretical results presented in lecture.
- *Mechanical skills*—To help students attain dexterity in handling apparatus.
- *Device experience*—To familiarize students with measuring tools.
- *Understanding Error*—To help students learn the tools of experiment as a method to convince others of your results: statistics, error analysis, and the ideas of accuracy and precision.
- *Concept building*—To help students understand fundamental physics concepts.
- *Empiricism*—To help students understand the empirical basis of science.
- *Exposure to research*—To help students get a feel for what scientific exploration and research are like.
- *Attitudes and expectations*—To help students build their understanding of the role of independent thought and coherence in scientific thinking.

This is a powerful and daunting list. Most laboratories have at first rather limited and practical goals—to satisfy the requirements of the engineering school or to qualify for premedical certification. In implementation, most laboratories only explicitly try to achieve the first two or three goals. Sometimes understanding error is an explicit goal, but in my experience, traditional laboratories fail badly at this goal. Students go through the motions but rarely understand the point. Extensive research on this issue is badly needed.

Often less happens in traditional labs than we might hope

In my research group's observation of traditional laboratories, one result is clear: the dialogs that take place are extremely narrow [Lippmann 2002]. Our videotapes show students spending most of the period trying to read the manual and figure out what it wants them to do. The students make little or no attempt to synthesize in order to get an overview of what the point of the lab is. Almost all of the discussion concentrates on the concrete questions of how to configure, run, and get information from the apparatus. There is little or no discussion of the purpose of the measurement, how it will be used, the physics to be extracted, or the limitations of the measurements. The students are so focused on achieving the "paper" goals of the lab—getting numbers to be able to construct lab reports—that the learning goals appear totally lost.

Of course, one might hope that they "get the numbers" in the lab and then "think about them" outside of class. This may be the case, but I suspect it is a pious hope. Students rarely have the skills to think deeply about experiments. This is where they need the help and guidance of an instructor, and they don't get it if this activity is carried out outside of class (or with instructors who can't handle the serious pedagogy needed).

A more interactive approach to the traditional laboratory

Some of my colleagues and TAs have experimented with variations in the traditional laboratory in order to get the students more intellectually engaged. From this anecdotal evidence, I extract a few tentative guidelines. I sincerely hope that in a few years, educational research will be able to "put legs" under these speculations.

- *Make it a "class" through discussion*—Often students in lab speak only to their lab partner. There is little sharing of results or problems with other students in the class. An overall class discussion at the beginning and end of the class might increase the engagement.
- *Take away the lab manual*—Having a step-by-step procedure may guarantee that most students complete the lab but undermines important learning goals. Pat Cooney at Millersville University has had success with simply writing the task on the board and having students figure out what they have to do.
- *Start the class with a planning discussion*—Most students do not spontaneously relate the broad goals of the lab to the details of the measurements. Having them think about these issues before beginning their measurements is probably a good idea. Bob Chambers at the University of Arizona has had good results with two-week labs in which the students use the first week to plan the experiment and the second week to carry it out.
- *Occasionally ask them what they're doing and why*—Students frequently get lost in the details of an activity and can get off on the wrong track. Asking them perspective questions ("What are you doing here? What will that tell you? What could go wrong?") might help them make the connection to the purpose of the experiment.
- *Share results*—Arranging labs so that there is some time for discussion and sharing of results at the end of the lab might help identify problems and give students a better idea of the meaning of experimental uncertainty.

The laboratory is the traditional instructional environment that is, in principle, best set up for independent active-engagement learning in line with our cognitive model of learning. Much more research will need to be done in order to figure out what learning goals can be effectively accomplished in the laboratory environment and how.

REALTIME PHYSICS

Environment: Laboratory.

Staff: One trained facilitator per 30 students.

Population: Introductory calculus-based physics students.

Computers: One for every two to four students.

Other Equipment: An analog-to-digital converter (ADC). Probes needed for the ADC include motion detector (sonic ranger), force probes, pressure and temperature probes, current and voltage probes, and rotary motion probe. Low-friction carts and tracks required for the mechanics experiments.

Time Investment: Low to moderate.

Available Materials: Three published manuals of laboratory worksheets for Mechanics (12 labs), Heat and Thermodynamics (6 labs), and electric circuits (8 labs) [Sokoloff 1995]. Laboratories in electricity and optics are under development. An instructor's guide is available on-line to registered instructors at http://www.wiley.com/college/sokoloff-physics/.

Sokoloff, Thornton, and Laws have recently combined to develop a new series of mechanics laboratories that can be used in a traditional lecture/lab/recitation teaching environment. They make heavy use of computer-assisted data acquisition and the results of research on student difficulties.

RTP uses cognitive conflict and technology to build concepts

The primary goal for these laboratory exercises is to help students acquire a good understanding of a set of related physics concepts [Thornton 1996]. Additional goals include providing students with experience using microcomputers for data collection, display, and analysis, and enhancing laboratory skills. The primary goal has been extensively tested by the designers and by other researchers using standardized evaluation surveys. Significant gains appear to be possible. (These results are discussed in detail at the end of this section.)

The critical tool in these laboratories is an analog-to-digital converter (ADC) connected to a computer. Many different probes can connect to these ADCs and provide the student with graphs of a wide variety of measured and inferred variables. Our senses do not provide us with direct measures of many of the quantities that are critical to an understanding of fundamental physics concepts. Our brains easily infer position and change in position, but inferring speed from visual data seems to be a learned skill (and one most people who have experience crossing streets learn effectively). Acceleration, on the other hand, seems to be quite a bit more difficult. Our brains easily infer rate of heat flow to the skin but are hard pressed to distinguish that from temperature. The computer probes allow live-time plots[13] of position, temperature, pressure, force, current, voltage, and even of such complex calculated

[13]"Live-time" in this context means that there is no noticeable time delay between the event being measured and the display of the data.

Figure 8.10 Analog-to-digital converters from Vernier and Pasco. Either box connects to the computer's serial port. A variety of probes can be connected to the box's front (shown).

variables as velocity, acceleration, and kinetic energy. The ADCs from Pasco and Vernier are shown in Figure 8.10.

The authors combine pedagogically validated methods such as cognitive conflict, bridging, and the learning cycle (exploration/concept introduction/concept application) with the power of computer-assisted data acquisition to help students re-map their interpretation of their experience with the physical world.[14]

RTP relies on psychological calibration of technology

When I first told Sagredo about the microcomputer use in the laboratory, he complained. "But if they don't understand how a measurement is made, they don't really understand what it means." This may be true, Sagredo. I am fairly certain that my introductory students who use the sonic ranger to measure velocity understand neither how the sounds are created and detected nor how the position data is transformed into velocity data. (They do seem to understand the idea that the sound's travel time gives a measure of distance—at least qualitatively.) On the other hand, I don't require that they know how their calculator calculates the sine function before I permit them to use it. When I have them carry out some indirect activity, such as producing a spark tape from some motion, making measurements of the positions, and then calculating and graphing the result, I may be helping them to understand how to make measurements, but the time delay between the motion itself and the production of the graph may be 15 minutes or more. This is far too long for them to buffer or rehearse their memory of the motion and make an intuitive connection.

A good example of how this works is the first RTP activity, "Introduction to Motion." This is the first lab and begins by having students use a sonic ranger motion detector (see Figure 8.11) to create position graphs of their own motions.[15] They see how the apparatus works by performing a series of constant-velocity motions and seeing what position graph appears. I refer to this process as a *psychological calibration.* They quickly get the idea and identify some interesting and relevant experimental issues, such as getting out of range of the beam of sound waves, getting too close (the motion detector only works at distances greater than about 50 cm), and seeing "bumps" produced by their individual steps. They then make predictions for a specific motion described in words, do the measurement, and reconcile any

[14] See chapter 2 and the discussion of primitives and facets.

[15] The ranger works be emitting clicks from a speaker in the ranger and measuring the time until an echo returns and is detected by a microphone in the ranger.

Figure 8.11 A sonic ranger motion detector from Vernier software.

discrepancies. Finally, they carry out a variety of position graphs in order to see a broader range of possibilities.

The experiment then turns to a study of velocity graphs. Again, they begin with a psychological calibration, measuring simple constant-velocity motions, inspecting the graphs, and comparing them to the position graphs. They then undertake an interesting activity: matching a given velocity graph, the one displayed in Figure 8.12. The graph is cleverly chosen. An initial velocity of 0.5 m/s away from the detector for 4 seconds produces a displacement of

In this activity, you will try to move to match a velocity—time graph shown on the computer screen. This is often much harder than matching a position graph as you did in the previous investigation. Most people find it quite a challenge at first to move so as to match a velocity graph. In fact, some velocity graphs that can be invented cannot be matched!

1. Open the experiment file called **Velocity Match (L1A2-2)** to display the velocity—time graph shown below on the screen.

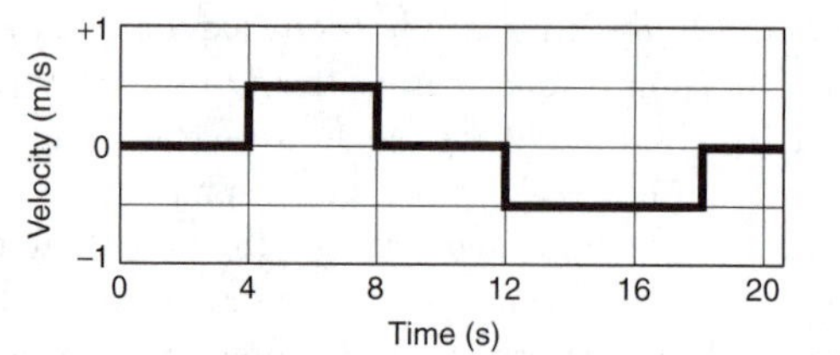

Prediction 2-2: Describe in words how you would move so that your velocity matched each part of this velocity-time graph.

2. **Begin graphing**, and move so as to eliminate this graph. You may try a number of times. Work as a team and plan your movements. Get the times right. Get the velocities right. Each person should take a turn. Draw in your group's best match on the axes above.

Question 2-4: Describe how you moved to match each part of the graph. Did this agree with your predictions?

Question 2-5: Is it possible for an object to move so that it produces an absolutely vertical line on a velocity—time graph? Explain.

Question 2-6: Did you run into the motion detector on your return trip? If so, why did this happen? How did you solve the problem? Does a velocity graph tell you where to start? Explain.

Figure 8.12 An activity from a RealTime Physics lab. Students are making use of a sonic ranger attached to a computer to display velocity graphs in live time.

2 m. A return velocity of –0.5 m/s for 6 seconds produces a displacement of −3 m. Although I have not run these laboratories at the University of Maryland, we have adapted them for the Activity-Based Physics Tutorials. In their prediction, students describe the motion required in terms of the velocities—speeds and directions—but rarely think about distances. (I've never seen anyone do it.) As a result, they start at the minimum distance from the ranger and begin by walking backward. After having gone back 2 meters, they try to come forward 3 meters and run into the detector. In their attempt to resolve the difficulty (sometimes a carefully placed question from the facilitator is needed), they effectively explore the relation between a velocity graph and the resulting displacements. They are then asked to predict position graphs given a velocity graph.

The RealTime Physics laboratory continues with an exploration of average velocity and with fitting the graph with a straight line using computer tools that are provided. Specific questions try to ensure that the students actually think about the result the computer is giving and do not simply take it as a given. Finally, the lab ends with a measurement of the velocity of a cart on a track in preparation for the second experiment, which is concerned with acceleration.

The above example illustrates many of the features common in RTP labs. They often include:

1. Psychological calibration of the measuring apparatus
2. Qualitative kinesthetic experiments (using one's own body as the object measured)
3. Predictions
4. Cognitive conflict
5. Representation translation
6. Quantitative measurement
7. Modeling data mathematically

RTP labs are effective in building concepts

The psychological calibration, live-time graphs, and kinesthetic experiments appear to have a powerful effect on intuition building. In my experience, students who have done these experiments are much more inclined to make physical sense of velocity and acceleration than students who have done more traditional experiments. Thornton and Sokoloff report data at their home institutions on subsets of the FMCE[16] with RTP laboratories [Thornton 1996]. At Tufts, Thornton tried the RTP laboratories with an off-semester calculus-based class of about 100 students. He reports that off-semester students tend to be less well prepared than those who begin physics immediately in the fall. The results are shown in Table 8.3. The gains were excellent. (The two rows report a cluster of questions probing Newton's first and second laws in a natural language [*n*] environment[17] and in a graphical [*g*] environment—the velocity graph questions.)

These results are compelling. The fractional gains achieved by the students with RTP are exceptional: greater than 0.8. Sagredo is skeptical. He points out that there is no direct

[16] See the discussion of the FMCE in chapter 5. The FMCE is on the Resource CD associated with this volume.

[17] These are the questions on the FMCE involving the sled. See the FMCE on the Resource CD.

TABLE 8.3 Pre- and Post-Results for Results on the FMCE's Sled [*n*] and Velocity Graph Questions [*g*] in Different Environments.

	Tufts RTP (pre)	Tufts RTP (post)	⟨g⟩	Oregon NOLAB (pre)	Oregon NOLAB (post)	⟨g⟩	Oregon RTP (pre)	Oregon RTP (post)	⟨g⟩
NI&II [n]	34%	92%	0.81	16%	22%	0.07	17%	82%	0.78
NI&II [g]	21%	94%	0.92	9%	15%	0.07	8%	83%	0.82

"head-to-head" comparison with a traditional laboratory and that the instruction was carried out at the primary institution. Moreover, the test used was developed by the researchers who developed the instruction. Sagredo is concerned that the extra hours of instruction the students had in laboratory (compared to those with no laboratory) might have made a big difference and that the instructors might have "taught to the test."

On the first issue, Sagredo, I'm not so concerned. It's my sense that traditional laboratories contribute little or nothing to conceptual learning. In fact, instructors in traditional courses making their best effort rarely produce fractional gains better than 0.2 to 0.35. Gains of 0.8 or 0.9 suggest that the method is highly effective.

The second item is of more concern. Of course, in a sense we *always* "teach to the test." As discussed in chapter 5, if the test is a good one, it measures what we want the students to learn. The problem occurs when instructors, knowing the wording of the questions to be used in an evaluation, focus—perhaps inadvertently—on cues that can lead students to recognize the physics that needs to be accessed for a particular question.

To deal with this issue, we have to see how well the instructional method "travels." The RTP/ILD dissemination project (supported by FIPSE) sponsored the implementation of RTP

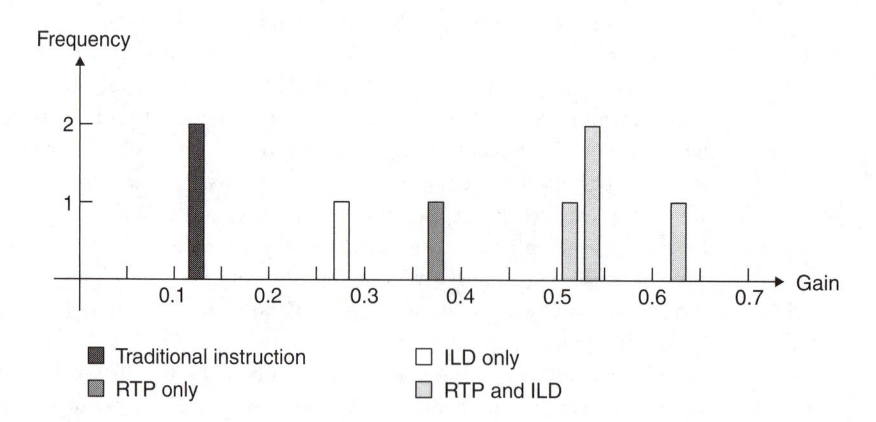

Figure 8.13 Fractional gains on the FMCE in five different colleges and universities that used fully traditional instruction, or traditional lectures with RTP, or RTP and ILDs ($N = 1000$) [Wittmann 2001].

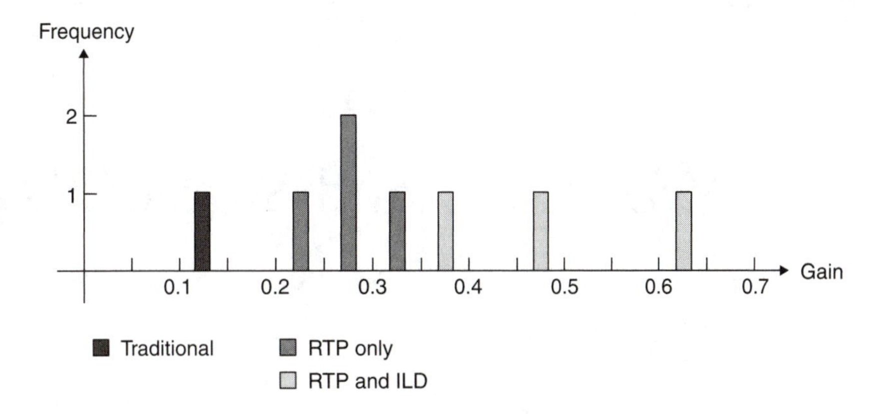

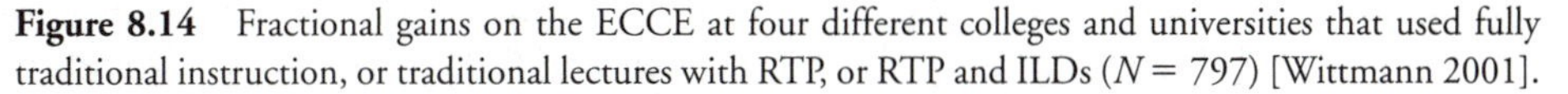

Figure 8.14 Fractional gains on the ECCE at four different colleges and universities that used fully traditional instruction, or traditional lectures with RTP, or RTP and ILDs ($N = 797$) [Wittmann 2001].

laboratories and Interactive Lecture Demonstrations (ILDs—see chapter 7) at a number of different colleges and universities. Preliminary results using the FMCE as a pre- and post-test show that the RTP Mechanics lab alone produces substantial improvement compared to schools using traditional laboratories [Wittmann 2001]. When the RTP Mechanics laboratories are supported by the use of ILDs in lecture, the results tend to be even better. These results are displayed in Figure 8.13.

I infer that the RTP laboratories can have a significant effect on student understanding of basic concepts and on their ability to use a variety of representations in thinking about these concepts. They tend to spend less time on error analysis than traditional labs, but since it is my sense that students in traditional labs typically "go through the motions" of error analysis but gain little real understanding ([Allie 1998] [Sere 1993]), this may be a case of giving up two birds in the bush in order to get one in the hand.

A preliminary version of the RTP Electricity labs was available for the FIPSE dissemination project. A preliminary analysis of the first-year implementation of these labs at secondary institutions using the Electric Circuits Concept Evaluation (ECCE)[18] pre and post shows strong gains with RTP over traditional and even stronger gains when RTP Electricity labs are supported by the use of ILDs in lecture (see Figure 8.14).

[18] The ECCE is on the Resource CD associated with this volume.

CHAPTER 9

Workshop and Studio Methods

I had become thoroughly disillusioned
by the ineffectiveness of the large general lecture courses
of which I had seen so much in Europe and also in Columbia,
and felt that a collegiate course in which laboratory problems
and assigned quiz problems carried the thread of the course
could be made to yield much better training, at least in physics.
I started with the idea of making the whole course self-contained . . .
I abolished the general lectures.
This general method of teaching . . . has been followed
in all the courses with which I have been in any way connected since.
Robert A. Millikan [Millikan 1950]

The Millikan quote in the epigraph shows that dissatisfaction with traditional lectures is not a new story. Although Millikan's Autobiography, from which the quote is taken, was published in 1950, the course he is describing was introduced in the first decade of the twentieth century. Many physicists, myself included, have the strong intuition that the empirical component in physics is a critical element and one that introductory students often fail to appreciate. This failure may occur in part because traditional lectures tend to be a series of didactic statements of "discovered truth" followed by complex mathematical derivations. An occasional ex-post-facto demonstration or laboratory experiment "to demonstrate the truth of the theoretical result presented in lecture" does little to help the student understand the fundamental grounding and development of physical ideas and principles in careful observation.

There are clearly many possible ways of remedying this oversight. Lectures could begin with the phenomena and build up the concepts as part of a need to describe a set of phenomena. Laws could be built from observed systematics in the behavior of physical systems. Laboratories could be of the guided discovery type and could introduce the material before lecture.

But perhaps the most dramatic modification of an introductory physics course is to adopt Millikan's method, in which "laboratory problems and assigned quiz problems carried the thread of the course." In the modern era, this approach has been developed under the rubric of *workshop* or *studio* courses, courses in which lecture plays a small (or nonexistent) role. All the class hours are combined into blocks of time in which the students work with laboratory equipment for most of the period.

Perhaps the first modern incarnation of this approach is *Physics by Inquiry* (PbI), a course for pre- and in-service teachers developed by Lillian McDermott and her collaborators at the University of Washington over the past 25 years [McDermott 1996]. In this class, there are no lectures at all. Students work through building the ideas of topics in physics using carefully guided laboratory manuals and simple equipment. Although PbI is explicitly designed for pre-service teachers and other nonscience majors, it is deep and rich enough that many of the lessons provide valuable ideas for the development of lessons even for calculus-based physics.

The PbI method was adapted for calculus-based physics in the late 1980s by Priscilla Laws of Dickinson College under the name *Workshop Physics* (WP). Since problem solving and developing quantitative experimental skills are goals not shared by the pre-service teacher class, Laws expanded McDermott's vision to include substantial components of modern computer-based laboratory tools, including computer-assisted data acquisition and data acquisition from video. (She and her collaborators developed many of these tools themselves.)

As set up at Dickinson, Workshop Physics runs in classes of 25 to 30 students. This is possible at a small liberal arts college like Dickinson where few students take introductory calculus-based physics.[1] Research-based institutions with engineering schools might have as many as 1000 students taking calculus-based physics in any particular term. Two attempts to bring something like WP to environments with large numbers of students occurred in the 1990s at Rensselaer Polytechnic Institute (*Studio Physics*) [Wilson 1992] [Wilson 1994] and North Carolina State University (*SCALE-UP*). The latter is described as a case study in chapter 10.

PHYSICS BY INQUIRY

Environment: Workshop.

Staff: One trained facilitator per 10–15 students.[2]

Population: Pre- and in-service K-12 teachers; underprepared students; nonscience majors.

Computers: Limited use.

Other Equipment: An extensive list of traditional laboratory equipment.

Time investment: Large.

Available Materials: A two-volume activity guide [McDermott 1996]. The Washington group runs a summer workshop to help interested instructors learn the approach.[3]

[1] Though the numbers grew substantially after the introduction of WP requiring the creation of multiple sections.

[2] In a remarkable experiment, the course has been taught with reasonable success using a single experienced instructor for 70 students [Scherr 2003].

[3] A video (*Physics by Inquiry: A Video Resource*) is available that provides illustrative examples of the materials being used. Contact the UWPEG for information.

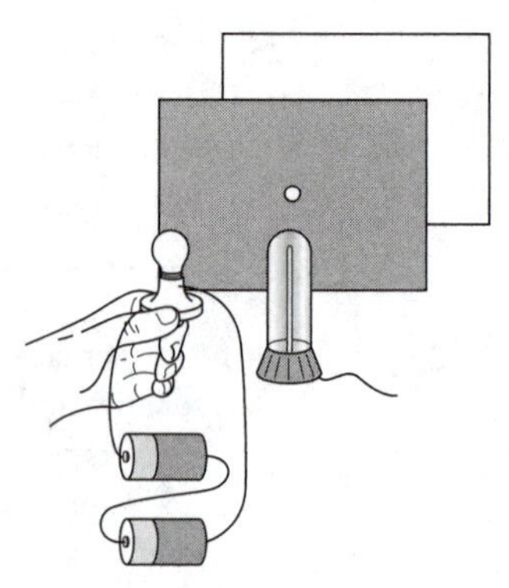

Figure 9.1 A simple apparatus from Physics by Inquiry.

One of the earlier modern prototypes of a full studio course was *Physics by Inquiry* (PbI), developed by Lillian McDermott and her colleagues at the University of Washington [McDermott 1996]. The course was developed for students studying to be teachers (*pre-service teachers* in the American terminology) and is a full guided-discovery laboratory. There is no lecture; students meet for three laboratory periods of two hours each per week. During these periods, students work in pairs with simple equipment and are guided to reason through physical examples with simple apparatus and carefully prepared worksheets. A sample apparatus for the unit on light is shown in Figure 9.1.

In PbI, students learn a few topics deeply

An assumption built into the material is that it is more important for the students to learn a few topics deeply and to build a sense of how the methods of science lead to "sense-making" about the physical world than to cover a large number of topics superficially. The materials emphasize specific concepts and specific elements of scientific reasoning such as control of variables and the use of multiple representations. The material is structured into independent modules (see Table 9.1), so a one- or multisemester term can be built by selecting two to three units per term. This has the advantage that if one permutes the choice of modules in successive years, in-service teachers can return to take the class in multiple terms without repeating material.

The worksheets are based on research in student understanding[4] and often use the cognitive conflict model in the *elicit/confront/resolve* form described in the discussion of Tutorials in chapter 8. The worksheets guide the students through observing physical phenomena, constructing hypotheses to explain the phenomena, and the testing of those hypotheses in new experiments. Trained facilitators (approximately one for every 10 to 15 students) help students to find their own path to understanding by guiding them with carefully chosen questions. Specific places are indicated in the lessons called *checkouts.* Students are instructed to check their results with a facilitator at this point before going on.

[4]Surprisingly, the Washington group has published very little of their research that has gone into the construction of Physics by Inquiry. Much of the group's published work on Tutorials (see references in [McDermott 1999] contained in the Appendix) on qualitative reasoning carries implications for PbI, despite the difference in populations.

TABLE 9.1 Modules in Physics by Inquiry

Volume I	Volume II
• Properties of Matter	• Electric Circuits
• Heat and Temperature	• Electromagnets
• Light and Color	• Light and Optics
• Magnets	• Kinematics
• Astronomy by Sight: The Sun, Moon, and Stars	• Astronomy by Sight: The Earth and the Solar System

During my sabbatical at the University of Washington (1992–1993), I participated in facilitating PbI classes. I was particularly impressed by the activity in the Astronomy module in which students made their own observations of the phase of the Moon and its position relative to the Sun over the entire term. Near the end of the term, the class's data were collected and discussed, and a model for how the Moon was lit was developed. Many students were surprised that they could see the Moon in the daytime, and many believed that the phases were caused by the Earth's shadow—a belief they could not sustain in light of the evidence. I myself realized for the first time that I could tell directions from the phase and position of the Moon, even after sunset.

Students may need help in changing their expectations for PbI

Physics by Inquiry is quite challenging for many students (even physics graduate students), as the goals, the structure of the learning environment, and the activities expected of the student differ dramatically from those they have learned to expect in traditional science classes. Some students at first resent the idea that they are not being given answers to memorize but that they have to work them out for themselves and have to understand how the laws and principles are supported by experiment. Students can exert considerable pressure on an instructor to change this. Careful facilitation is needed throughout the course to help students pay attention to what they are supposed to be doing—thinking, reasoning, and making sense of what they see in a coherent and consistent fashion. The first few weeks of a PbI class can be quite tumultuous, but it is worth riding out the storm. The Washington group offers both extended summer workshops in Seattle and short workshops at meetings of the American Association of Physics Teachers to help would-be PbI-ers learn the ropes.

Evaluations of PbI show it to be very effective

Although there is not a large body of published literature on the success of PbI, the observations of a few researchers on secondary implementations of PbI are worth mentioning.

In a recent paper, Lillian McDermott and her colleagues reported on a secondary implementation of PbI for pre-service elementary school teachers at the University of Cyprus [McDermott 2000].[5] They evaluated the performance of students on direct current circuits

[5] The classes that used PbI used it in Greek translation.

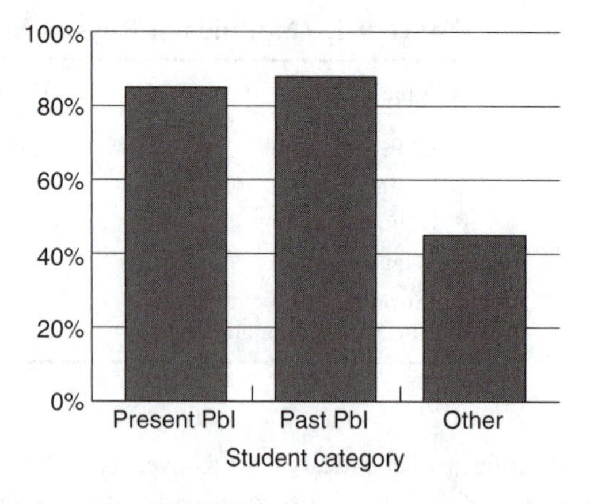

Figure 9.2 Post-test results on the DIRECT concept survey given to students at the University of Cyprus recently completing PbI, completing PbI in the previous year, and recently completing a more traditional constructivist physics course for teachers [McDermott 2000].

using the DIRECT conceptual test of understanding of DC circuits. (The DIRECT survey is on the Resource CD associated with this volume.) Three groups of students were compared: 102 students who had just completed the electric circuits module of the PbI course; a group of 102 students who completed the module in the previous year; and a group of 101 students who had just completed the topic in a course using constructivist pedagogy but not using the findings of discipline-based research or the research-redevelopment cycle. (See Figure 6.1.) The results are shown in Figure 9.2.

Beth Thacker and her colleagues at the Ohio State University compared student success on a qualitative circuits problem in her secondary implementation of a PbI class with the same pair of problems given to engineering physics students, students in an honors physics class, and a traditional physics class for nonscience majors [Thacker 1994]. One problem they referred to as a synthesis problem. It required only qualitative reasoning. (See Figure 9.3.) A second problem they referred to as an analysis problem. It required quantitative (algebraic, not numeric) reasoning. (See Figure 9.4.) The instructor in the engineering class thought that the problem was exactly appropriate for his students and that they should have little difficulty with it. The instructor in the honors physics class thought the problem was too easy but was willing to give it as extra credit.

The PbI students scored significantly better than either of the other groups on the synthesis (qualitative) problem and significantly better than the engineers on the analysis problem. (See Figure 9.5.) Note that an answer was not considered to be correct unless the student gave an explanation that included a reason. A restatement of the result (e.g., "Bulb D is unaffected.") was not considered sufficient.

A preliminary study of the Ohio State PbI students using the MPEX showed significant gains on the concept variable [May 2000].

(Synthesis A). All of the bulbs in the figures below have the same resistance R. If bulb B is removed from the circuit, what happens to the current through (brightness of) bulb A, bulb D and the battery? Indicate whether it increases, decreases, or remains the same. Explain your reasoning.

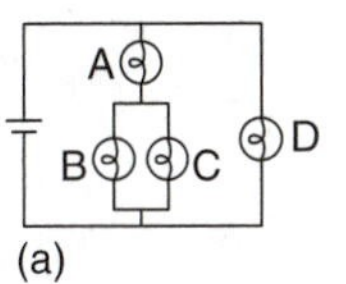

(a)

(Synthesis B). A wire is added to the circuit in the figure below. What happens to the current through (brightness of) bulb A, bulb D and the battery? Indicate whether it increases, decreases, or remains the same. Explain your reasoning.

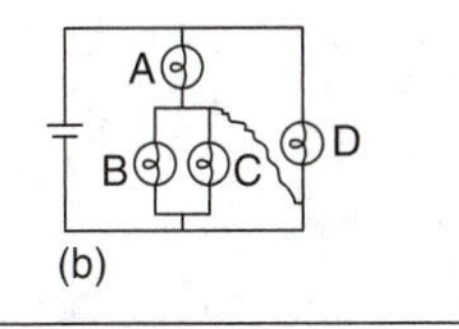

(b)

Figure 9.3 A qualitative reasoning (synthesis) circuits problem given to test Physics by Inquiry students [Thacker 1994].

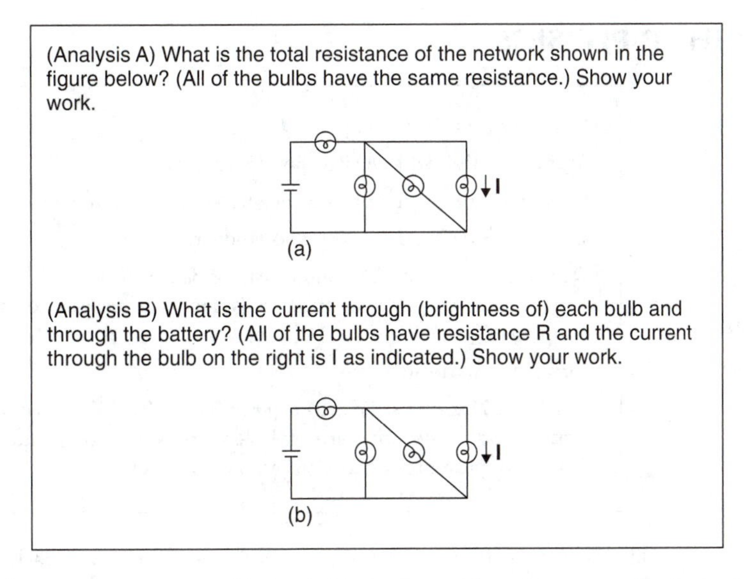

Figure 9.4 A quantitative reasoning (analysis) circuits problem given to test Physics by Inquiry students [Thacker 1994].

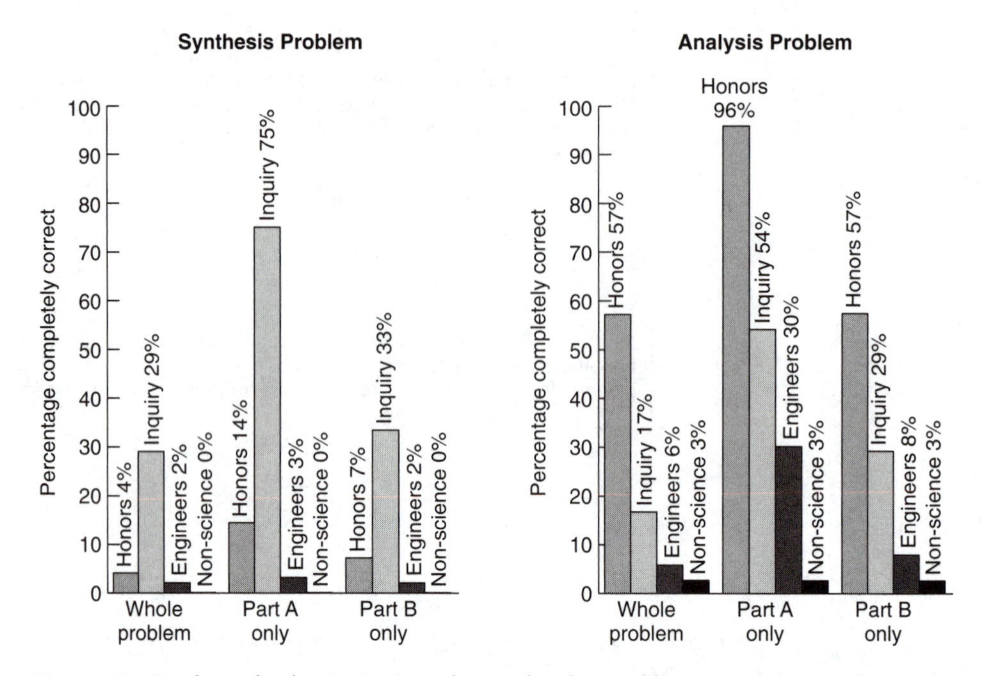

Figure 9.5 Results on the electric circuit synthesis and analysis problems given to honors physics, PbI, engineering physics, and physics for nonscience classes [Thacker 1994].

WORKSHOP PHYSICS

Environment: Workshop.

Staff: One trained facilitator per 15 students.

Population: Introductory calculus-based physics students.

Computers: One for every two students.

Other Equipment: Computer-assisted data acquisition devices (ADCs) and probes, spreadsheet software, *Videopoint*™ (video data analysis tools), standard laboratory equipment.

Time investment: Large.

Available Materials: An activity guide [Laws 1999]. Extensive sets of homework problems and other resources are available at the WP website: http://physics.dickinson.edu/ A listserve promotes discussions among WP users.

The Workshop Physics (WP) class was developed by Priscilla Laws and her collaborators at Dickinson College [Laws 1991] [Laws 1999] using the research-redevelopment cycle discussed in chapter 6. In the mid-1980s, Laws became deeply involved in the use of the computer in the laboratory, developing laboratory tools for working with Atari computers. In the late

1980s, Laws and Ron Thornton of Tufts University, working with a number of fine young programmers, developed a "stable platform" for microcomputer-based laboratory activities. The Universal Laboratory Interface box (ULI) is an analog-to-digital converter.[6] One end connects to the computer's serial port and the other to a "shoebox full" of probes—motion detectors, force probes, temperature sensors, pressure gauges, voltage probes, and so on. (The ADCs from Vernier and Pasco are shown in Figure 8.10. The Vernier motion detector is shown in Figure 8.11.) Software, available for both Wintel and Mac environments, allows the students to display graphs of any measured variables against any others, to fit the graphs with various mathematical functions, to read values off the curves, to integrate the curves between chosen limits, and so on. Spreadsheets (and perhaps symbol manipulators) provide the students with tools for mathematical modeling of their experimental results.

Students in WP build their concepts using technology

What it is the students actually do in this class is hinted at by the structure of the classroom, shown in Figure 6.4. The students function in groups as in the inquiry-style classroom, each pair working with a computer workstation with the computer-assisted data collection structure and modeling tools described above. Classes are held in three two-hour periods per week. During these classes, most of the student time is spent with apparatus—making observations and building mathematical models of their results. The classroom contains a central area for common demonstrations, and many class periods may include brief lecture segments or whole-class discussions.

Students are guided through the process of carrying out, making sense of, and modeling their experiments with worksheets contained in an Activity Guide [Laws 1999]. In addition to the Activity Guide, students are assigned reading in a text and homework problems. Although the homework may include traditional end-of-chapter problems, the WP group has developed a series of context-rich problems, many of which use video or other computer-collected data. An example is given in Figure 9.7.

Figure 9.6 Computer-assisted data-acquisition setup showing, from right to left, computer, Vernier ULI and motion detector, PASCO cart on track.

[6] The design of this box was based on previous devices developed by Bob Tinker and his colleagues at TERC.

In the early spring of 1995, the quarterly Radcliffe College Alumnae magazine featured a cover story of working women in the 1990s. The unusual cover depicts a young woman pushing up on a glass ceiling. Assume that she is not moving. Give three reasons why the woman's position in this photo is impossible. What physical laws or principles are being violated? How do you think this photograph might have been made?

Figure 9.7 A sample of a context-rich Workshop Physics problem.

WP is developed through and informed by education research

Although the Dickinson group focuses on development rather than on basic educational research, the development of the Workshop Physics materials relies heavily on published physics education research and on careful local observations using the research-redevelopment cycle. An excellent example of how this works is given on the WP web pages. Upon reading the research papers on student difficulty with direct current circuits published by McDermott and Shaffer [McDermott 1992] [Shaffer 1992], Laws began modifying her WP materials on the subject. She evaluated students' conceptual learning on the topic using the ECCE developed by Sokoloff and Thornton and included on the Resource CD associated with this volume. She compared her results with those obtained by Sokoloff at the University of Oregon after students received traditional lectures on the topic. Pre-tests at both Dickinson and Oregon showed that students entered the class with little knowledge of the subject, missing about 70% of the questions. Lectures helped little, reducing the error rates to about 65%. Students in Workshop Physics did substantially better, attaining average error rates of as low as 40%. However, after reading the McDermott-Shaffer papers, Laws redesigned the WP activities. The results were a substantial improvement, with error rates falling to less than 10%. These results are shown in Figure 9.8. Similar results are displayed on the WP website http://www.physics.dickinson.edu for topics in kinematics, dynamics, and thermodynamics.

WP changes the frame in which students work

Implementing Workshop Physics can be a nontrivial activity as the workshop-style class may violate a number of student expectations. Students who come to a physics class expecting a

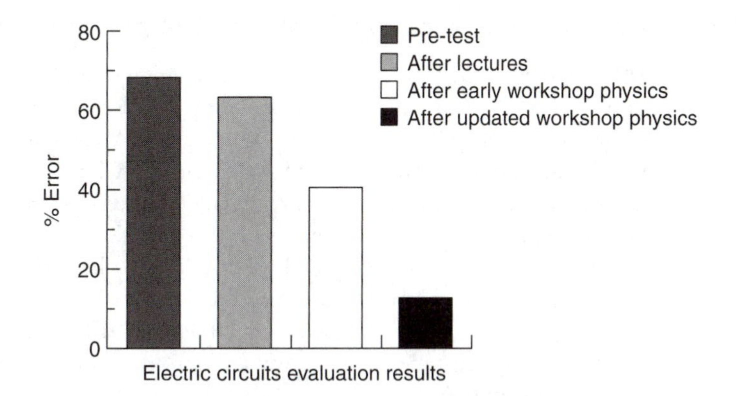

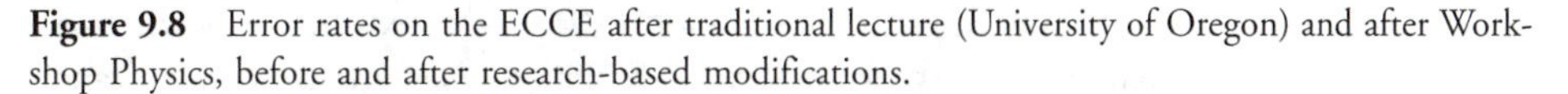

Figure 9.8 Error rates on the ECCE after traditional lecture (University of Oregon) and after Workshop Physics, before and after research-based modifications.

lecture and lots of plug-and-chug homework problems may be dismayed by the amount of thinking involved. Students who have had high school physics may expect their physics to be math-dominated rather than experiment-dominated. And students who are unaccustomed to group work may have trouble interacting appropriately.

Workshop Physics is an attempt to seriously change the framework of learning to have students focus more strongly on understanding and on the experimental basis of the physics. Getting students to understand not just the physics but how to make this shift of mental frame can be difficult. Implementing a course like Workshop Physics effectively requires that the instructor be sensitive to all these complex issues and be aware of the need to renegotiate the instructor–student social contract.

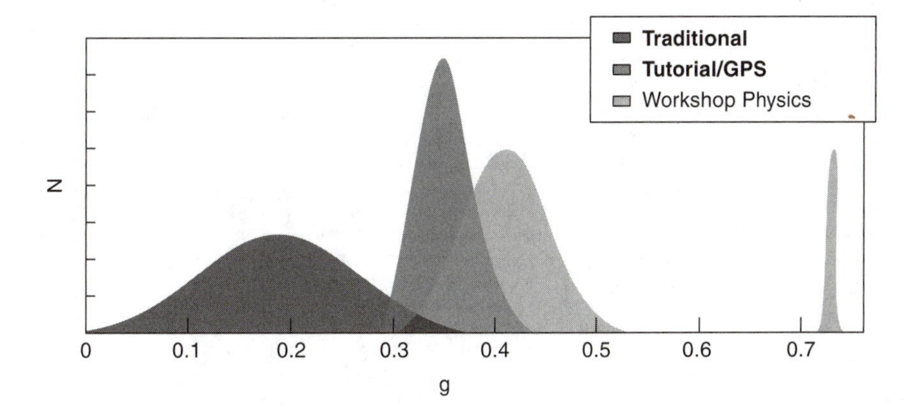

Figure 9.9 Distribution of fractional gains on pre-post FCI and FMCE for traditional, recitation modifications (Tutorial and CGPS), and Workshop Physics. The histograms for each group are fit with a normalized Gaussian. The spike at the right corresponds to WP at Dickinson College.

Evaluations of WP show it to be highly effective in building concepts

Jeff Saul and I carried out an independent evaluation of student learning in Workshop Physics as part of the WP dissemination project (supported by FIPSE) [Saul 1997]. Our study included seven colleges and universities implementing Workshop Physics for the first or second time. Student learning was evaluated with pre-post FCI or FMCE, with common exam questions, and through interviews with 27 student volunteers at three of the dissemination schools. Student expectations were measured with the MPEX.

The results from the pre-post FCI/FMCE are schematically shown in Figure 9.9. The secondary WP implementation averaged fractional gains of 0.41 ± 0.02 (SEM) compared to 0.20 ± 0.03 for the traditional classes and 0.34 ± 0.01 for the recitation modifications. (The mature primary implementation of WP at Dickinson College typically attains fractional gains on these tests of 0.74.)

MPEX averages in the traditional classes showed the pre-post deterioration described in chapter 5. The early secondary implementations showed no significant loss and occasional small gains on the reality link measure. WP at Dickinson College shows significant gains on the cognitive cluster of independence-coherence-concepts.

CHAPTER 10

Using the Physics Suite

To suppose that scientific findings decide
the value of educational undertakings is to reverse the real case.
Actual activities in educating test the worth of the results
of scientific results. They may be scientific in some other field,
but not in education until they serve educational purposes,
and whether they really serve or not can be found out only in practice.
John Dewey [Dewey 1929]

We've come a long way since I first introduced the idea of the Physics Suite at the beginning of chapter 1. In subsequent chapters, we talked about some of what is known from research about student thinking and learning, and I described some innovations in curriculum development based on that research. Some of those innovations belong to the Physics Suite, while the others can be adapted to work with Suite elements. We are now ready to revisit the elements of the Suite to consider how you might use them in your teaching.

The materials of the Physics Suite have been set up so that you can either (1) use many of them at the same time, or (2) integrate one or more elements with the materials you are already using. The Suite is not a radical change to the traditional approach to introductory physics teaching. It is meant to provide elements that are both familiar and improved as a result of what has been learned from physics education research and as a result of new developments in educational technology. You can choose to incrementally adopt individual elements of the Suite that are appropriate in your instructional environment.

In this chapter, I begin with a review of the principles behind the Physics Suite. This is followed by a brief recap of the elements of the Suite, along with ideas for using the Physics Suite in different environments. I conclude by presenting four case studies that give specific examples of how various instructors have adopted and adapted elements of the Suite in high schools, colleges, and universities.

THE PRINCIPLES BEHIND THE PHYSICS SUITE

The Physics Suite is predicated on two shifts of perspective from the traditional approach to teaching, one sociological, one psychological.

First, over the years, physics instructors (myself included) have typically assumed that they should present the content in a manner that satisfies themselves and that the students would then take responsibility for doing whatever they needed to do to learn the material. This results in a filter that passes only those students who come to the class with the drive, motivation, and understanding of the study skills necessary to succeed in physics on their own. This turns out to be a small fraction of the students who take physics. With the increasing shift in the emphasis of physics instruction to a service course preparing scientists and engineers who will not necessarily become physicists, we need to shift our assumptions. Now we want to see how much help we can offer students who need to know some physics but who may not know how to learn physics appropriately.

Second, a lot has been learned from educational and psychological research that can help instructors understand how to help students learn how to learn physics. The critical principle is:[1]

> *Principle 1:* Individuals build their knowledge by making connections to existing knowledge; they use this knowledge by productively creating a response to the information they receive.

The implication of this principle is that what matters most in a course is what the students actually do. In order to have effective instruction, we, therefore, have to create learning environments that encourage and enable students to do what they need to do to learn, even if they don"t choose (or know how) to do so spontaneously.

This result is often called upon to justify the creation of "hands-on" or "active learning" environments. Unfortunately, this largely misses the point. Students can actively work with equipment and still not learn very much physics. (See, for example, the discussion of traditional laboratories in chapter 9.) What matters is their pattern of thought as they blend hands-on activities with reflection. We need a more detailed understanding of student learning in order to be able to design environments that effectively encourage appropriate thought and reflection. Further complicating the situation is what we need to do to achieve our goals depends on our goals. And these goals depend on a both external and internal factors: the population we teaching, the course and its nominal purpose, our own individual goals as instructors,[2] and our model of student thinking and learning.

Our model of thinking and learning is often implicit, tacit, and in contradiction with both fundamental research in cognitive psychology and the observed behavior of students in educational situations. In chapter 2, I have put together a "soft paradigm" (a set of guidelines or heuristics) that can help instructors develop and apply a more sophisticated approach to thinking and learning. The fundamental ideas are that long-term memory is productive, associative, and context dependent. Principles 2–5 in chapter 2 (the context, change, individid-

[1]See chapter 2 for details.

[2]For example, in teaching our engineering physics class, I very much want my students to learn to understand and create arguments based on symmetry. Sagredo wants them to appreciate physics as a creative historical evolution of ideas. Both are noble and justifiable goals but are not required in the classes we teach.

uality, and social learning principles) help us to understand what features of an environment might be appropriate to help students create appropriate understandings.

> *Principle 2:* What people construct depends on the context—including their mental states.
>
> *Principle 3:* It is reasonably easy to learn something that matches or extends an existing schema, but changing a well-established schema substantially is difficult.
>
> *Principle 4:* Since each individual constructs his or her own mental structures, different students have different mental responses and different approaches to learning. Any population of students will show a significant variation in a large number of cognitive variables.
>
> *Principle 5:* For most individuals, learning is most effectively carried out via social interactions.

This model has a number of implications. Principle 1 implies that it helps if we pay careful attention to what students know and how they use that knowledge in creating both their correct and incorrect understandings of what we are trying to teach them. This is the resource component of the model. Principle 2 suggests that it helps if we pay careful attention to *when* students access the knowledge we want them to have. This means teaching them how to use it effectively and to recognize when it is appropriate. This is the linking component of the model. Principle 3 reminds us of the importance of knowing "what they know that ain't so" and provides guidance in building environments that help students get on the right track. The bridging and cognitive conflict approaches discussed in chapter 2 provide two possible approaches. Principle 4 reminds us to provide environments appropriate for a variety of student styles. This is the diversity component of the model. Finally, Principle 5 suggests we pay attention to the design of the social environments in which our students learn. This is the social component of the model.

Chapter 2 also discusses and develops how this model leads us to articulate a particular collection of explicit goals for our physics instruction. Goals that have been considered in the construction of the Physics Suite include:

> *Goal 1: Concepts*—Our students should understand what the physics they are learning is about in terms of a strong base in concepts firmly rooted in the physical world.
>
> *Goal 2: Coherence*—Our students should link the knowledge they acquire in their physics class into coherent physical models.
>
> *Goal 3: Functionality*—Our students should learn both how to use the physics they are learning and when to use it.

The Physics Suite is designed in accordance with this model of thinking and learning so as to help teachers create learning environments that function effectively.

THE ELEMENTS OF THE PHYSICS SUITE

Traditional instructional materials are organized around a content list and a textbook. The Physics Suite is designed to help instructors refocus their courses on learning goals and student activities. The text in the Physics Suite is intended to be supportive, but it is just one component in an array of materials that can help build an effective learning environment. (See Figure 1.2 for a diagram showing the elements of the Suite.)

I briefly describe each element in turn and show how it fits into the overall picture. Since the modifications to the narrative—an important Suite element—are not discussed elsewhere in this volume, I consider these in more detail than the rest.

The Suite's narrative text: Understanding Physics

The narrative for the Physics Suite, *Understanding Physics* (Cummings, Laws, Cooney, and Redish) [Cumming 2003], is based on the sixth edition of the popular *Fundamentals of Physics* (Halliday, Resnick, and Walker) [HRW6 2001]. It is being adapted and modified to an active-learning environment in a number of ways.

1. *Modifications to the text are incremental, not radical.* The next generation of texts may integrate activities directly and be presented on-line. *Understanding Physics* begins with a standard text and takes a step in that direction. Existing in-text activities (Reading Exercises and Sample Problems) are enhanced, and links are indicated where connections to activities are appropriate and relevant. But it still looks enough like a traditional text to be in the "comfort zone" of both students and teachers accustomed to traditional texts. The primary active-learning enhancements come from working with additional Suite elements.

2. *The text is modified to take student difficulties into account.* At the time of this writing, thousands of papers have reported on the difficulties students have in learning physics [Pfundt 1994]. These are summarized in the Resource Letter on the Resource CD associated with this volume [McDermott 1999], and the results are discussed in a number of texts and instructor's guides [Arons 1990] [Reif 1995] [Viennot 2001] [Knight 2002]. In *Understanding Physics*, issues that are well known to confuse students and cause them difficulty are discussed with care. Many traditional texts consider these issues trivial and brush them off with a sentence or use ill-chosen examples that may actually activate classic misconceptions.

3. *Topics are introduced by making connections to personal experience whenever possible and appropriate.* We try to follow the principle "idea first, name later" and to motivate a discussion before it occurs, making contact with the student's personal experience. This helps build patterns of association between the physics students are learning and the knowledge they already have, and helps them reinterpret their experiences in a way that is consistent with physical laws.

4. *Material is explained in a logical order.* We try to follow the "given-new" principle (see chapter 2) and build on ideas the student can be expected to understand, based on the resources they bring from their everyday experiences. Many texts present results didactically, starting a discussion by stating a complex result at a point where a student does not have the resources to understand or interpret it and then explaining it through a complex exposition taking many pages.

5. *Concepts are emphasized.* One of the primary goals in our model is that students make sense of the physics they are learning. This is impossible if they see physics as a set of abstruse equations. We therefore stress conceptual and qualitative understanding from the first and continually make connections between equations and conceptual ideas.

6. *Reading exercises and sample (touchstone) problems are limited and carefully chosen.* In an effort to provide examples and items of interest, physics texts often include large numbers of text boxes, sidebars, and sample problems. This can make the narrative choppy and difficult for the student to follow. In *Understanding Physics,* reading exercises are carefully selected

to provide appropriate thinking and reflection activities at the end of a section. These are often suitable (especially in small classes) as topics for discussion. Sample problems have been transformed to "touchstones"—carefully chosen examples that illustrate key points to help students understand how to use the physics within a problem. Sample problems that only illustrated straightforward equation application and manipulation ("plug-and-chug") have been removed.

7. *Examples and illustrations often use familiar computer tools.* Examples in the text have been expanded and modified to use computer-assisted data acquisition and analysis (CADAA) tools and collection of data from video. Other elements of the Suite—laboratories, tutorials, interactive lecture demonstrations, and Workshop Physics—make heavy use of this technology as well. This has a number of advantages. It connects the text to the student's experiences in other parts of the class; it connects directly to real-world experiences (through video); it uses realistic rather than idealized data; and it connects the narrative to the more active Suite elements.

8. *No chapter summaries are provided.* This is a feature, not a bug! We didn't forget to include the chapter summaries; rather, we removed them intentionally. Students tend to use pre-created summaries as a crutch to grab equations for plug-and-chug purposes and as a shortcut to avoid reading (and trying to make sense of) the text. Providing an "authority-validated" summary in the text both robs the students of the opportunity to construct summaries for themselves and sends the covert message to trust authority instead of building their own judgment. Instructors who feel that summaries are essential (as do I) can assign students to create summaries as a regular part of their written homework.

9. *The order of materials has been modified somewhat to be more pedagogically coherent.* Some of the traditional orderings emphasize the mathematical structure of the material at the expense of the physics or violate the given-new principle. For example, free fall is often included in the kinematics chapter, since constant acceleration problems can be solved algebraically. One result of this approach is that students are often confused by gravity, being unable to disentangle the idea of the gravitational field near the Earth's surface ($g = 9.8$ N/kg) from the gravitational acceleration that results in free fall ($a_g = 9.8$ m/s^2). We treat free fall in chapter 3 after a discussion of force. In order to emphasize the centrality of Newtonian dynamics, Newton's second law is treated in one dimension immediately after the definitions of velocity and acceleration. Momentum is treated as a natural extension of Newton's second law. The concept of energy is delayed until after the discussion of extended objects.

10. *Vector mathematics is handled in a just-in-time fashion.* The dynamics of one-dimensional motion is presented before introducing general two- and three-dimensional vectors. Vectors and vector products are introduced as they are needed, with the dot product being presented in association with the concept of work and the cross product being presented in association with the concept of torque. One-dimensional motion is presented in the context of one-dimensional vectors, with a notation that is consistent with general vector notation to help alleviate a traditional confusion students have between scalars and vector components.

Using the Suite in lab: RealTime Physics

RealTime Physics (RTP) is a set of three published laboratory modules covering the topics Mechanics (12 labs), Heat and Thermodynamics (6 labs), and Electric Circuits (8 labs).[3]

[3]A module on Light and Optics is currently under development.

These labs help students build a good understanding of fundamental concepts through use of a guided inquiry model with cognitive conflict. Experiments rely heavily on computer-assisted data acquisition to enable students to collect high-quality data quickly and easily. This allows students to perform many experiments and to focus on phenomena rather than on data taking. Initial activities with new probes help students "psychologically calibrate" the probes, that is, convince themselves that they understand what the probes' responses mean, even though they may not be clear on how the probes produce their data. Research shows that these labs can be very effective in helping students build concepts. For a more detailed discussion, see chapter 8.

Implementing RealTime Physics requires a laboratory setup with computer-assisted data acquisition equipment for every two or three students.

Using the Suite in lecture: Interactive Lecture Demonstrations

Interactive Lecture Demonstrations (ILDs) help students learn representation-translation skills and strengthen their conceptual understanding through active engagement in a large lecture environment. Students receive two copies of a worksheet: one for making predictions and one for summarizing observations. The instructor goes through a sequence of carefully chosen demonstrations using computer-assisted data acquisition to display graphs of results on a large screen in real time. Students are shown the demonstration without data collection. They are then given the opportunity to make predictions and to discuss their predictions with their neighbors before the results are collected and displayed. The topics and demonstrations rely heavily on research that identifies common misconceptions and difficulties. The worksheets use cognitive conflict and social learning. Research shows that these activities can be very effective in helping students both learn concepts and understand graphical representations. They can also be effective in smaller classes. For a more detailed discussion, see chapter 7.

Implementing Interactive Lecture Demonstrations requires only a single computer (for the lecturer) with computer-assisted data acquisition and a large-screen display. It takes a bit of practice for a traditional lecturer to develop the interactive style that gets students contributing to the discussion in a way that makes these demonstrations most effective. (See chapter 7.)

Using the Suite in recitation sections: Tutorials

Tutorials are a curricular environment for delivering active conceptual development in recitation sections. They have a tight, carefully guided group-learning structure similar in feel to the RealTime Physics labs or the Interactive Lecture Demonstrations. They are based on research on student difficulties and make frequent use of both cognitive conflict and bridging. An extensive set of Tutorials has been developed by the University of Washington Physics Education Group covering a wide range of topics from kinematics to physical optics [Tutorials 1998]. These Tutorials are designed to be usable in environments without computer tools, so they make almost no use of computer-assisted data acquisition or video. A supplementary set of tutorials using computer technology including computer-assisted data acquisition, video display and analysis, and simulations, are available as part of the Suite [ABP Tutorials]. Tutorials have been shown to be effective in improving concept learning compared to classes with traditional recitations. For a more detailed discussion, see chapter 8.

Implementing the UWPEG Tutorials requires some small items consisting of standard physics laboratory equipment and inexpensive materials from a hardware store. Implementing the ABP Tutorials requires a computer and data acquisition tools for every three to four students. Both types of Tutorials require approximately one facilitator per 15 students. These facilitators need training, both to make sure they understand the physics (which can be quite subtle and challenging, even for faculty and graduate students in physics) and to help them learn a "semi-Socratic" approach, in which the instructor guides with a few well-chosen questions instead of explanations.

Putting it all together: Workshop Physics

Workshop Physics (WP) is the most radical component of the Physics Suite. It presumes a complete structural change from the traditional lecture/recitation/lab pattern. Typically, the class is structured into three two-hour laboratory sessions in which the students use sophisticated technology to build their physics knowledge through observation and mathematical modeling. Classes move smoothly back and forth from brief lecture segments, to class discussions, to full-class demonstrations, to small-group experimenting and modeling. An integrated set of computer tools are used for data acquisition, video capture and analysis, and graphing and modeling with spreadsheets.[4] Workshop Physics is extremely effective in classes of 30 or fewer, but it is difficult to deliver to hundreds of students. (See, however, the discussions of the North Carolina State case study below.) For a more detailed discussion of WP, see chapter 9.

Implementing Workshop Physics requires computer equipment, including a variety of data acquisition probes and tool software. One facilitator for every 15 students or so is a must, but if an instructor is present, they can include peer instructors (students who have successfully completed the course in a previous term). Learning to manage the laboratory logistics and to help students shift the expectations they might have developed in high school or other science courses can be a challenge and may take a few semesters before things run smoothly, but the gains both in learning and in student attitudes can be dramatic.

Homework and exams: Problems and questions

As discussed in chapter 4, the problems students solve, both for homework and on exams, are a critical part of the activities students carry out to learn physics. The choice of exam problems is particularly important, since exams send students both overt and covert messages about what they are supposed to be learning in class (whether we intend to send those messages or not). Traditional courses often limit homework or exams to questions that have numerical or multiple-choice answers so as to be easy to grade. This has the impact of undermining any more sophisticated messages we might send in other parts of the course about the richness of learning and thinking about physics and the value of learning to make sense of a physics problem. The Physics Suite includes an enhanced array of problems for homework and exams, including estimation problems, open-ended reasoning problems, context-rich problems, and essay questions.

Implementing more open homework and exam problems requires some structure for grading. Students need the feedback and motivation that grading provides in order for them

[4]These tools are contained on the Resource CD associated with this volume.

to take more complex and open-ended problems with the required degree of seriousness and reflection. This requires someone—an instructor or assistant—to spend some time evaluating questions. This can be difficult in large classes, but grading a small number of such questions (two to three per week, one to two per exam) can have a big impact.

Evaluating instruction: The Action Research Kit

As discussed in chapter 5, over the past 20 years, physics education research has documented student difficulties in a wide variety of topics in introductory physics. Using these results, researchers have constructed standardized conceptual surveys. Many of the items in these surveys are well designed. They focus on critical issues that are difficult for many students and they have attractive distractors that correspond to common student misconceptions.[5] Because of the strong context dependence in the response of novice students, these surveys (and especially a small number of items extracted from a survey) do not necessarily provide a good measure of an individual student's knowledge. A broader test with many contexts is required for that. But these surveys do give some idea of how much a class has learned, especially when given before and after instruction. More than a dozen surveys are provided on the Resource CD associated with this volume.

Suite compatible elements

Three non-Suite elements that can be comfortably used in conjunction with other Suite elements materials are Peer Instruction, Just-in-Time Teaching (JiTT), and Cooperative Problem Solving. These Suite-compatible materials are discussed in detail in chapters 7 and 8.

Peer Instruction

Peer Instruction is a method in which an instructor stops the class every 10 to 15 minutes to ask a challenging short-answer or multiple-choice question. Usually, the questions are qualitative and conceptual and activate a cognitive conflict for a significant number of students. The students choose an answer for themselves and then discuss it with a neighbor. Next, the results are collected (by raising hands, holding up cards, or via an electronic student response system), displayed, and reflected on in a whole-class discussion. Implementing Peer Instruction only requires a good set of closed-ended questions and problems. Choosing appropriate and effective problems is not easy. They must reflect a critical conceptual issue, a significant number of students (>20%) must get them wrong, and a significant number of students (>20%) must get them right. As with Interactive Lecture Demonstrations, learning to run a good Peer Instruction class discussion can take some practice. Mazur's book on the method contains a large number of potentially useful problems and helps in getting a good start with the approach [Mazur 1997]. For a more detailed discussion, see chapter 7.

Just-in-Time Teaching (JiTT)

JiTT is a method in which students respond to carefully constructed questions (including essay and context-rich questions) on-line. The instructor reviews the student answers before

[5] Recall from our discussion of the resource component of our learning model that a "misconception" does not necessarily refer to a stored "alternative theory." It may be produced on the spot by a student using spontaneous associations to inappropriate resources or inappropriate mappings of appropriate resources.

lecture and adapts the lecture to address student difficulties displayed in the answers, sometimes showing (anonymous) quotes for discussion. This method sends the valuable message that the instructor cares about whether students learn and is responding to them. A significant number of appropriately structured problems are contained in the book by Novak, Patterson, Gavrin, and Christian [Novak 1999]. For a more detailed discussion, see chapter 7.

Cooperative Problem Solving

Cooperative Problem Solving is a method for helping students learn to think about complex physics problems and solve them by working in groups of three in a recitation section or small class. The method employs heterogeneous grouping of students and assignment of roles, and it offers the students a structured method to learn to think about how to approach a complex problem. This method is very effective in helping students both develop good conceptual understanding and learn to solve problems. It sends the valuable message that one doesn't have to be able to see how to do a problem immediately in order to solve it, something many students at the introductory level fail to appreciate. A large number of useful problems are available on the website of the Minnesota group that developed the method.[6] For a more detailed discussion, see chapter 8.

All three of these methods are based on underlying cognitive models and goals similar to the Physics Suite and coordinate well with it.

USING THE PHYSICS SUITE IN DIFFERENT ENVIRONMENTS

To use parts of the Physics Suite effectively in your classroom requires two elements: a good match between the Suite elements chosen and the physical classroom conditions, and a good match between the philosophical orientation of the Suite and the orientation of the instructors involved.

Some of the Suite elements (RTP, WP, ABP Tutorials) rely heavily on student interactions with computers and computer-based laboratory equipment. Use of these elements requires approximately one computer station for every three to four students. These can, of course, be run in small sections in parallel. One laboratory with 8 to 10 computer stations can easily serve 400 to 500 students in the course of a week. Some elements of the Suite (RTP, WP, Tutorials) require facilitators—one instructor for every 15 students in the classroom. In these environments, students struggle in small groups with ideas and concepts. They require frequent (but not too frequent) checking, coaching, and guiding. In principle, one instructor with considerable experience in the methods can handle 30 students (or more), but it is difficult to pull off. A summary of these physical constraints is given in Table 10.1.

The role of room layout

The room layout plays an important role in using some of the Suite elements effectively. It is difficult to get students working together effectively in a lecture hall whose chairs are all oriented in one direction and bolted down. It is difficult to interact effectively with students working in a computer laboratory in which students sit individually or in pairs at computers facing in one direction (toward an assumed lecturer) and bolted down in rows. In these kinds of computer rooms, performing laboratory experiments is nearly impossible. Effective

[6]http://www.physics.umn.edu/groups/physed/

TABLE 10.1 Suite Elements Appropriate for a Variety of Environments.

Element	Large Classes $(S/F > 50)$	Small Classes $(S/F < 50)$	Facilitator Support $(S/F < 20)$	No Facilitator Support	Computer Rich $(S/C < 3)$	Computer Poor $(N \approx 1)$
Text (UP)	✓	✓	✓	✓	✓	✓
Lab (RTP)	✓	✓	✓		✓	
Lecture (ILD)	✓	✓	✓	✓	✓	✓
Recitation (UW Tutorials)	✓	✓	✓		✓	✓
Recitation (ABP Tutorials)	✓	✓	✓		✓	
Workshop Physics (WP)		✓	✓		✓	

room layouts for using various elements of the Suite are discussed in chapters 6 through 9. These layouts give students the opportunity for face-to-face interaction in small groups.

The role of facilitators

Another important consideration is that the facilitators have the appropriate philosophy and approach, and that they know how to listen to students and respond appropriately. Despite the best of intentions, this may not be easy. I had been teaching for 20 years before I realized that when students asked me questions, I was responding as a student rather than as a teacher. Having been a student for 20 years, having been rewarded for giving good answers to teachers' questions, and having been successful at getting those rewards, I had a very strong tendency to try to give the best answer I could to any question posed. Once I realized (embarrassingly late in my teaching career) that the point was not getting the question answered correctly but getting the student to learn and understand, I shifted my strategy.

Now, instead of answering students' questions directly, I try to diagnosis their real problem. What do they know that they can build an understanding on? What are they confused or wrong about that is going to cause them trouble? As a result, instead of answering a question right off, I ask some questions back. Often, I discover that students are trying to hide a confusion by creating questions that sound as if they know what they are talking about. Helping them to finding resources within themselves that they can bring to bear often makes all the difference. ("Oh! You mean it's like . . .") Even after 10 years of operating in this new mode, I still detect a strong tendency to want to give "a good answer," and sometimes, I even talk myself into believing that for some students, in some situations, it's appropriate.

Peer and graduate student facilitators may find it particularly hard to be in the right interactive mode. They are still students and tend to easily fall into the mode they use in answering their teachers' questions. When we first began testing Tutorials at Maryland in the mid-1990s, my department helped out by letting me handpick some of our best TAs. This turned out to be a problem. These TAs had developed their reputation by being articulate explainers. Often in that first semester, I had to pry them out from inside a group of four students where they had just spent 10 minutes, with pencil in hand, "showing" the students the answers to all the tutorial questions while the students sat watching, silently.

Finding the right balance of questions and answers, of intervention and "benign neglect," is difficult. The balance depends on so many things—the particular students involved, the task, the set of expectations that have been negotiated between student and instructor, and how tired or frustrated the students are. The key in making the gestalt shift from good student to effective teacher is learning to listen to the students and to consider them, as well as the content being discussed.

The large variety of materials offered in the Physics Suite along with the set of Suite-compatible materials offer instructors a large range of options. Instructors in different environments can use the materials in different ways.

FOUR CASE STUDIES: ADOPTING AND ADAPTING SUITE ELEMENTS

Every high school, college, and university physics class is a unique environment. Each has its own population of students, its own physical environment, its own history of teaching, its own faculty, and its own relations with other parts of its institution. Any implementation of new instructional materials must be adapted to each institution's unique characteristics and constraints. To illustrate how this plays out in real-world situations, in this section, I present four case studies of different kinds of institutions that have implemented various elements of the Physics Suite. The first two cases concern one or a few individuals teaching reasonably small classes: a public high school and a small liberal arts college. The second two concern large research universities that teach many students: one without and one with a physics education research group. These stories are based on interviews with some of the faculty involved in implementing the materials, on examination of their materials, and on data from their websites. I particularly want to thank Maxine Willis, Juliet Brosing, Mary Fehrs, Gary Gladding, Bob Beichner, and Jeff Saul for discussions.

Using Suite elements at a small institution

Gettysburg High School

Gettysburg High School (GHS) is a medium-sized high school in rural Pennsylvania. It serves a county that covers 185 square miles and has about 25,000 people. GHS has about 1200 students in four grades. The population draws from a wide demographic, ranging from the children of professional suburbanites to children who live in rural poverty and who will be the first generation in their family to attend college.

One of the teachers at Gettysburg, Maxine Willis, has been adapting her class to new developments in physics instruction over the past 15 years. She now uses many Suite elements, including *Understanding Physics* (UP), Workshop Physics (WP), Interactive Lecture Demonstrations (ILDs), RealTime Physics (RTP), and the WP Tools.

Maxine teaches both a standard physics class (noncalculus) and an AP physics class.[7] Typically, she teaches 40 to 50 students in standard physics divided into two sections and about 30 students in AP physics, again divided into two sections. The classes are therefore reasonably small and amenable to highly interactive environments with extended class discussions.

Class periods at GHS are 40 minutes long. Physics is taught in double periods five days a week to allow for lab work. Once a week, each class also meets in a single period for problem solving, answering questions, and recitation-like discussions. Since they are using double periods, they complete a standard one-year high school physics course in one semester.

[7]This is equivalent to a calculus-based university course in mechanics in preparation for AP Physics C.

Maxine's classroom is arranged as a Workshop Physics room (see Figure 6.4) and seats up to 28 students. The room has 14 computers for students, plus one for the instructor. The instructor's computer is connected to a flat-plate overhead-projector LCD panel. She is typically able to get peer instructors for each AP class—students who have previously completed the class successfully and who get independent study credit for their participation. According to Table 10.1, this makes GHS a small class with facilitator support and a computer-rich environment. They are therefore able to use all the elements of the Suite.

Maxine has been working with Suite elements in their various development stages since about 1989. By now she has considerable experience with them and can use them flexibly and creatively. In the AP class, she uses the text, extensive Workshop Physics activities and tools, ILDs, and the problem solution book. In the standard class, she uses activities selected for Workshop Physics, RealTime Physics labs (sometimes substituting the simpler Tools for Scientific Thinking labs), and ILDs.

In the AP class, two typical days might include a WP activity and a problem-solving activity. On a WP day, Maxine may begin by explaining some features of the equipment the students need to understand to carry out the task, but most of the period is spent with the students carrying out the activities themselves. Maxine and her peer assistant wander the classroom, asking and answering questions (and "answers" are often guiding questions). If there is time left at the end of the period, they may have a reflective discussion of what has been learned. Otherwise, that discussion takes place at the beginning of the next class.

Typically, WP activities are concept building. Maxine begins a topic with these and doesn't turn to serious problem solving until she feels that her students are clear on the concepts. If WP is not working for them, and in some cases where a WP activity is too complex for high school or uses too much equipment, she will substitute an ILD, taking a full double period to complete it.

On a problem-solving day, the students are supposed to have attempted some homework problems chosen from the text before coming to class.[8] They divide into groups of two to three. Each group is given a piece of whiteboard ($2' \times 2'$) and markers and is assigned a group number. Maxine then passes out the solution manual, and the students check their answers against the solutions in the manual. While they are doing this, she writes the problem numbers on the board. When a group decides that they have had difficulty with a problem, they put their group's attempt at a solution on the board under the problem number. Both the instructor and the class can then see the pattern of difficulties. If the entire class has had difficulty with a problem, Maxine will do a similar example (not the same problem). She then selects the problems most of the class had difficulty with and has the groups work them out on their whiteboards.

There are two critical elements in this activity. First, the students have a pattern they have to follow—they are required to include a diagram and show their line of reasoning. Second, the solutions in the manual usually are incomplete. The solutions in the manual pay little attention to the problem setup and tend to focus on the algebraic manipulations.[9] As a result, the solutions provide hints but don't fill in the critical thinking steps; the students have to do that themselves. By using the hints in the solution book and by working together,

[8] Students who fail to do this have their grades for the missed problems reduced.

[9] In this context, this is a feature, not a bug!

almost all students are able to work out and understand the solutions to all the problems. Typically, a class can do three or four problems per time block.

Maxine's experience with traditional texts in these classes has been poor. Students in the standard class can't make much sense of an introductory physics text, and even the AP students had considerable difficulty with earlier editions of Halliday, Resnik, and Walker. Students felt they couldn't understand it. She helped them by creating reading guides—questions to help them interpret what they were reading—especially in the first few chapters when they were getting started. Maxine reports that since adopting the preliminary edition of *Understanding Physics*, this problem has gone away. Her AP students are reading the text carefully and don't need explicit guidance. (*Understanding Physics* is not appropriate for her standard physics students since it uses calculus.)

Maxine reports that the elements of the Physics Suite work well for her and she is satisfied that student learning has improved—and not just for memorized facts. When I asked her what she thought the overall impact of adopting the Suite approach was, she said, "It's made me not the center of the classroom. The focus is more on the student as learner. More of my students are able to think physically. Once they know it, they have a really strong foundation. I feel like I'm giving future scientists their ABCs. I'm not covering a lot of material, but they're becoming much more powerful analytically than they were before. In addition, they're more confident problem solvers. They go off to competitive colleges and don't feel swamped anymore. This is a big improvement. Many of my students used to start out as science majors in college and then switch out. That doesn't happen nearly as often now."

Pacific University

Pacific University is a private college with professional graduate programs. The professional programs are mostly in the health sciences (occupational therapy, optometry, physician assistants, and psychology). Many of the undergraduates are interested in biology and in health science careers. Pacific is located in rural Oregon, in a small town up against the coastal range. It has about a thousand undergraduates and a thousand students in the professional programs.

The Physics Department is small—four faculty members plus one shared with optometry (3.3 FTEs). The Department teaches three introductory physics classes: conceptual physics, algebra-based physics, and calculus-based physics. Elements of the Physics Suite have been used in the latter two classes for a number of years.

The algebra-based class has become substantially smaller recently, since the Biology Department no longer requires it.[10] The number of students now fluctuates between 20 and 60. It is taught as a two-semester course meeting six hours/week as three one-hour lectures and one three-hour lab period. The three-hour period is split between tutorial instruction and laboratory.

The Pacific Physics Department has been using the RealTime Physics materials in their laboratories for about eight years. Most of the RTP labs are designed for three-hour blocks, so they have adapted them to their environment. Many of the RTP labs come in three parts, each appropriate for a one-hour period. They experimented with a variety of options and wound up choosing two of the three parts of an RTP lab and splitting it over two weeks. In

[10]Mary Fehrs reports that according to the biologists the students have to learn so much new biology that there is no room for courses that are of limited relevance.

each three-hour block, students do a two-hour tutorial and one hour of lab. (They found that their students were able to handle splitting the lab better than they were able to handle splitting tutorials.) The RTP labs follow the new mechanics sequence, which does not jibe with some standard texts. Rather than change the order of reading the text, they keep the text order and rearrange the order of the RTP labs to match. This seems to work OK.

Some course items (such as circuits) are taught in lab and are only mentioned briefly in lecture. The instructors at Pacific have chosen not to use the RTP pre-lab assignments since their students don't use these materials as a probe of their own thinking, as intended, but rather look the material up in the text to be sure of getting the right answers. This preempts the discovery character of the lab learning. The labs work for them without the students having completed the pre-lab materials. They do use the lab homeworks but modify them somewhat to fit the language and content of their lectures.

The instructors have struggled somewhat with the tutorial instruction. The UW Tutorials are designed for a calculus-based class and were too sophisticated for their population. Instructors have been using some problem-solving tutorials developed for the algebra-based class but are not satisfied with them and so are still hunting for appropriate materials that emphasize concept building. They do tutorials as a two-hour block in the three-hour period.

Mary Fehrs reports that in lecture she tries to put the material in context and tie it to concepts, and that she does some problem solving. She also does some ILDs and finds them very helpful. She reports that in ILDs her students can often be coaxed to make a thoughtful prediction—which they don't often do in lab or tutorial. They can be wrong and very confident about their wrong answers. Sometimes they have to see a result twice before they believe it. She tried using some JiTT over the web but gave it up because of difficulties managing their computer environment.

Mary's sense is that the algebra-based students are "good students"—that is, they will do just about whatever you ask of them. Unfortunately, their mode of successful learning up to this class has been to memorize and replay, and traditional lecturing plays right into that mode. The use of RTP and ILDs helps break this pattern and leads to considerable improvement. Mary says, "I've taught for 30 years. From daily walking and talking you get an idea of what they're getting. They get much more this way. They're really starting to think." The class results on FMCE show fractional gains of about 0.5—which is very good compared to the 0 to 0.3 found in a typical lecture-based environment [Wittmann 2001].

In the calculus-based class, the instructors adopt a full workshop model and use the Workshop Physics materials. Typically, they have about 20 students and a few peer instructors (students who have previously taken the course who are paid to facilitate during class and do grading). The physical setup consists of 1950s-style lab rooms with six long tables, set up with four students per table. Each table has a computer and an analog-to-digital converter for data acquisition. The students work in groups of twos, but the setup fosters interactions between two pairs. The classes meet for six hours/week in three two-hour blocks.

In this case, they use the Workshop Physics materials as is, without modification. They haven't used a textbook (though as of this writing they are planning to try the preliminary edition of *Understanding Physics*) and have used WP problems exclusively. They write new problems for exams but do not feel the need to create additional curricular materials. Mary Fehrs says, "[The WP materials] make a coherent whole and good conceptual sense as is."

Some of their exams include components that are laboratory oriented—analysis of data, work with spreadsheets, etc.

When I asked what problems she had encountered, Mary reported two problems with WP: the students have difficulty getting the "big picture" working with WP alone, and she has trouble getting the students to take their predictions seriously and "think hard" about them. In order to help students develop overviews, she has them write weekly summaries describing what they have learned during the week. She reports that this exercise helps them get perspective and organize the material somewhat, but she hopes that having a text will provide the perspective absent in the hands-on-oriented WP activity guide. She continues to work on finding ways to help students understand what she wants them to do in the prediction parts of the lesson. The results on pre-post testing for their WP students are very strong fractional gains—about 0.6 [Wittmann 2001].

Mary is quite satisfied with the use of WP for this class and says she would never go back to lecturing. Her other colleagues have bought in to the method, and she says they would expect any new hires to continue using the approach. She likes the fact that WP "immediately tells the students that learning is active, not passive." She accepts the fact that there is always a diverse response, with some students loving the approach and some hating it. She says, "It's a lot of work and we don't cover as many topics as we used to. But Workshop Physics doesn't let you fool yourself into thinking that your students understand something that they really don't. There is constant feedback that reminds you what they haven't learned yet. It's easy to fool yourself when you lecture."

Using Suite elements at a large institution

In both the case studies discussed above, a small number of instructors (one to three) were dealing with a reasonably small number of students (<100). A significant fraction (about a third) of the students in the United States taking physics in a service course at the college level do so in large public research universities. These universities may have between 10,000 and 45,000 students, departments with 15 to 75 faculty members, and graduate students to serve as TAs. Calculus-based physics may serve as many as 500 to 1000 students in each class in each term. Managing the laboratories, recitations, homework, and exams for an operation of this scale can be daunting. Because of the large number of students, large lectures seem inevitable, and often many different faculty members have responsibility for the same class. Although departmental committees often choose textbooks and content may be constrained,[11] faculty are often given considerable leeway in designing their approach to the class. Laboratories may be run independently from the lecture/recitation sections. These strongly held cultural constraints can make implementing lasting reform difficult.

Two large universities that have managed reform even within these constraints are the University of Illinois and North Carolina State University. Both are large engineering schools. The University of Illinois has adopted and adapted a number of Physics Suite elements within the context of the traditional large lecture/recitation/laboratory environment and has created its own

[11] A common textbook and constraints on content permit students to choose different faculty members' classes in different terms in order to handle shifts in the scheduling of their other classes.

web-homework tool. North Carolina State, with the help of an on-site physics education research group, has creatively adapted the workshop approach to a large class environment.

The University of Illinois

At the University of Illinois, Urbana-Champagne (UIUC), in the mid-1990s, the department head, David Campbell, convinced his colleagues that the results coming out of physics education research implied that their large-lecture traditional approach to introductory physics was not as effective as it could be. The motivations and first steps are described in his article "Parallel Parking an Aircraft Carrier: Revising the Calculus-Based Introductory Physics Sequence at Illinois," for the *Newsletter of the Forum on Education of the APS* [Campbell 1997].[12]

The UIUC is a large state university with a large high-quality, research-oriented physics department and is one of the premier engineering schools in the country. The Physics Department at UIUC offers three semesters of calculus-based physics and two semesters of algebra-based physics, teaching all classes every term. A total of about 2500 students register for these classes each semester. The large number of students requires a large number of faculty and a large infrastructure, including TAs and lab managers.

Before they reformed the program, the Physics Department at UIUC taught classes in a traditional fashion with lecture sections of 200 to 300 students for three hours/week and recitation and lab sections of 24 students for three to four hours per week. The lecturer was responsible for all aspects of lecture, recitation, and homework. The TAs planned their sections largely on their own and mostly answered questions on problem solving by demonstrating the solutions themselves at the board. Labs were in the standard "cookbook" model and were the responsibility of a faculty member who had little or no contact with the lecturers.

The result was that neither faculty nor students were happy. Faculty felt that managing a large lecture section with associated homework and TAs was a difficult and unrewarding experience. Students mostly expected to dislike physics and found that the course confirmed their expectations. In pre-post surveys of student attitudes (see Figure 10.1), more than half of the students said they considered physics "negative or awful," with the number increasing in the end-of-semester survey.

In 1995, the faculty agreed to participate in a major reform of the calculus-based physics class. Computers were available in labs, and graduate students were available to serve as facilitators. Available space included traditional large lecture halls with fixed tiered seating, small recitation classrooms with movable chairs, and traditional laboratory space with long tables. Funds were made available to provide computers and data acquisition devices for the laboratory. According to Table 10.1, this makes UIUC a large class with facilitator support and a computer-rich environment. Therefore all of the elements of the Suite can be used with the exception of Workshop Physics.

Since each of the three classes was taught each semester, the reforms had to be implemented "in flight." The schedule of implementation is shown in Figure 10.2.[13] They decided

[12] This article and the other articles in the FED newsletter are available on-line at http://www.aps.org/.

[13] The plan—figuring out what to do, preparing materials, and implementing the results—should be contrasted with the cyclic model displayed in Figure 6.1. The lack of a research-based cycle implies that corrections and updates have to be handled explicitly in some other way.

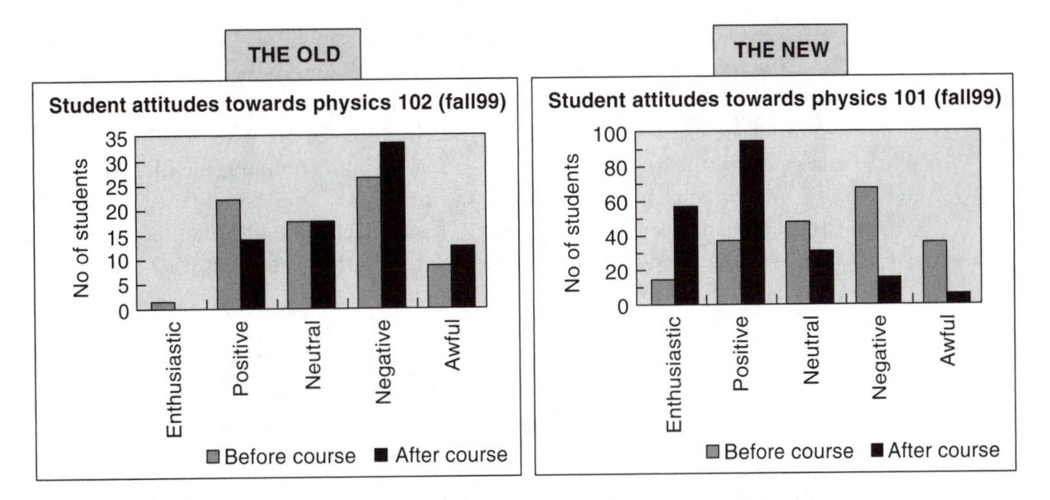

Figure 10.1 Results of pre-post attitude surveys at the University of Illinois—Urbana-Champagne, before and after curriculum reform. Courtesy of Gary Gladding [Gladding 2001].

to restructure the course to provide more active engagement activities for the students and to provide a more balanced load for the faculty. A primary design criterion was to produce a more coherent and integrated course—and one that would be seen as belonging to the department, not whose individual pieces belonged to individual faculty members or TAs.

The calculus-based course was restructured to include two 75-minute lectures, a two-hour recitation, and a two-hour laboratory each week. In the algebra-based course, they restructured to include two 50-minute lectures, a two-hour recitation, and a three-hour laboratory each week. Lecture classes were increased in size so that one of the faculty members formerly assigned to lecture could be assigned to manage the recitations and homework. This resulted in a more balanced teaching load. The decision was made to implement Peer Instruction in lectures, Tutorials and Cooperative Problem Solving (with home-grown

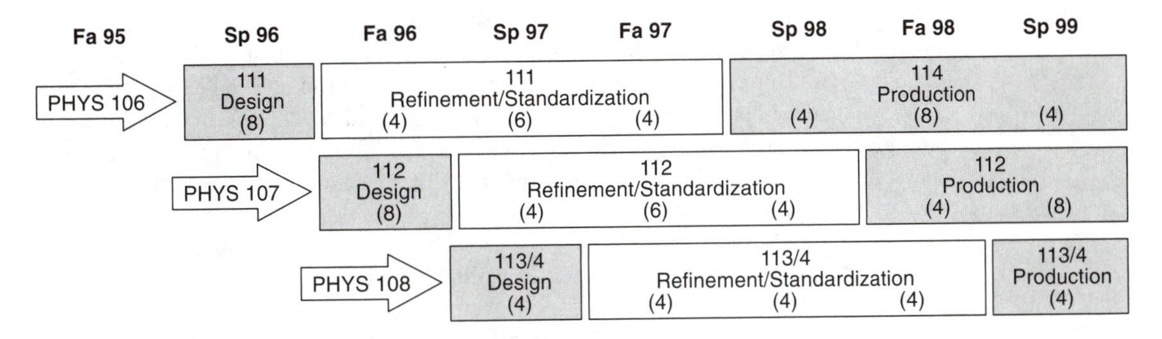

Figure 10.2 Design and implementation schedule used at the University of Illinois—Urbana-Champagne, to reform the calculus-based physics course [Gladding 2001].

problems) in recitation, and RealTime Physics laboratories.[14] In order to create coherence and a sense of common ownership, the working team meets regularly to discuss what is happening in the class. Problems for the cooperative-problem-solving sessions are written in common, as are exam problems. Exams are multiple-choice and machine-graded, and they are delivered in the evenings to all sections at the same time outside of class hours.

A new emphasis on concepts was introduced, and a web-based homework system was developed and maintained by Denny Kane, a full-time staff member. (See [Steltzer 2001].) The web-delivered homework comes in three different formats, each serving a specific pedagogical purpose.

1. Linked quantitative problems around a specific physical situation: Brief hints are available upon request, and immediate feedback (right or wrong) is given.
2. Delayed feedback homework: Similar in structure to the first, but no feedback is given until after the grading deadline. These are like on-line quizzes.
3. Interactive examples: A single multistep quantitative problem with extensive help dialogs.

The UIUC Physics Department made a substantial commitment of both funds and staff in order to implement the program. Existing positions of computer coordination and lecturer were repurposed for the new structure, and a new Associate Head position was created to manage the system.

Both students and faculty are enthusiastic about the results. On pre-post happiness surveys, ratings were dramatically improved. (See Fig. 10.1.) At the end of the first semester, more than 75% reported that they were either positive or enthusiastic about their physics class. (This compares with less than 20% before the reform.) Far more TAs made the campus's "list of excellent TAs" after reform than did before (77% compared to 19% before reforms). Faculty are much more comfortable with the teaching load than previously, and calculus-based physics is no longer considered a "killer" teaching assignment. More details can be obtained from the course website.[15]

North Carolina State University

The University of Illinois began their reforms with the premise that changing a large class involving thousands of students and dozens of faculty required maintaining a lecture-based format, and they adapted many research-based curriculum reforms that fit that model. North Carolina State University began their reforms with a different idea. The Physics Department at NC State contains a physics education research group led by Bob Beichner. Bob felt quite strongly that inquiry-based instructional models such as Workshop Physics could lead to dramatic improvements in learning over lecture-based models, so he set about trying to find a way to implement one.

NC State University, like UIUC, is a large state engineering school with a research-oriented physics department. The Physics Department has between 50 and 60 faculty members

[14] Algebra-based physics labs were adapted from pre-publication versions of RTP, and calculus-based labs were created on site using a predict-observe-explain model similar to RTP, ILDs, and Tutorials.

[15] http://www.physics.uiuc.edu/education/course_revision.html

and instructs more than 5000 students in physics classes each year.[16] The largest class is calculus-based physics, a two-semester course with about 500 to 1000 students in each class in each semester. The reform of the calculus-based introductory physics class was undertaken in quite a different way from the UIUC as a result of the presence of a PER group. The project was begun in the years 1995–1997 with a small class, observers and interviewers from the research group, and standardized survey instruments to measure student progress. In the initial phase of the project, all the students' classes (physics, calculus, chemistry, and introduction to engineering) were done in a coordinated fashion [Beichner 1999]. In later stages, the project developed a stand-alone method for physics referred to as Student-Centered Activities for Large Enrollment University Physics (SCALE-UP). The SCALE-UP project received funding for development and dissemination from U.S. government funding agencies and is currently being adapted at a number of other universities.[17]

In the initial stage of the project, the approach planned was described to entering engineering students, and they were asked to volunteer to participate. Approximately 10% of the students volunteered, and half of those were chosen at random to participate in the experimental class. The other half took the traditional class and were used as a control group.

The class was set up to operate in a workshop/studio mode, and material was adapted from a wide variety of research-based sources, including Workshop Physics, Physics by Inquiry, Cooperative Problem Solving, and Peer Instruction. Students were organized into groups of three heterogeneously and the same groups worked together in all their classes. Roles were assigned, and students received instruction both on how to work in groups and how to approach complex problems.

Large numbers of computers with data acquisition and modeling tools (spreadsheets and Interactive Physics) were available, as were graduate student facilitators. Experimentation with different layouts led them to select round tables with 9 to 12 students and 3 to 4 laptop computers. Before and after views of the physics classroom are shown in Figure 10.3.

Figure 10.3 Views of the physics classroom at NC State before and after the transformations created by the SCALE-UP project.

[16] Number of students enrolled in physics classes each year including summer sessions.

[17] See the NC State SCALE UP website at http://www2.ncsu.edu/ncsu/pams/physics/Physics_Ed/ for current information on the project.

TABLE 10.2 Fraction of Students Who Received Grades of C or Better in All Their Math, Chemistry, Physics, and Engineering Classes.

	1995–1996		1996–1997	
	N	Success Rate	N	Success Rate
Test class	35	69%	36	78%
Control group	31	52%		
Traditional	736	52%	552	50%

Classes are run workshop style, with an intermix of brief lecture elements, discussion, problem solving, laboratory investigations, and modeling. Laboratory segments are often brief—10 minutes or so (though they sometimes grow to blocks of as much as an hour or two)—and in response to questions raised in class discussion. UW Tutorials are used but are broken up into short discussion segments of 10 to 15 minutes. Since there are no formal lectures, students are responsible for reading materials before each class.

Like the reforms at UIUC, the reforms at NC State made significant use of the web, in particular, the *WebAssign* environment, developed and supported at NC State. The web is used for distribution of materials, maintaining the class schedule, and distributing and collecting homework. WebAssign is used both in and out of class to present questions and problems to the students. Use is also made of Java applets, particularly the Physlet collection [Christian 2001].

The classes were initially run with a class size of about 30. They were then increased to 54 and currently run successfully with as many as eight tables and 99 students in a room at once.

The results of the initial attempts showed significant success. The rate of good grades across the group of classes was much higher in the test group than among traditional students (and the control group was close to traditional). (See Table 10.2.)

Student success in physics learning was also better than in the traditional class. The average score on the TUG-K for the test class was 89% ± 2%, while for the traditional students it was 42% ± 2% (standard error of the mean). On the FCI pre-post, the test class had an average fractional gain of 0.42 ± 0.06 and 0.55 ± 0.05 in the two years reported. The control group only achieved an average fractional gain of 0.21 ± 0.04, comparable to the average reported for traditional classes [Hake 1992][Redish 1997]. On the shared midsemester exam, the test class did significantly better than the control group that received traditional instruction (80% to 68%).[18] Pre-post MPEX studies showed no change on most variables (a good result, considering that almost all classes show a significant loss) and a 1.5σ improvement on the coherence variable. On other attitudinal variables, students in the trial class showed substantial improvement in confidence levels, while students in the traditional class showed declines (especially those in the control group).[19]

[18] In the following semester, in which all students received traditional instruction in E&M (electricity and magnetism), no difference was noted between the groups of students.

[19] See the project's annual reports on the NC State website for more details.

As part of the SCALE-UP project, Beichner and his collaborators are building up a large collection of adapted and modified materials, including short hands-on activities, interesting questions to consider, and group-based laboratory exercises that require a lab report. Check the group's website for information on availability of these materials.[20]

CONCLUSION

As these case studies show, there are many paths to reform. The particular path you choose depends on the resources you have available, your constraints, and above all, the opportunities offered by your most important resources—the individuals in your department who show an interest in changing physics instruction at your institution. The Physics Suite offers you and your colleagues tools to work with in your efforts to improve what your students take from their physics classes.

[20] http://www2.ncsu.edu/ncsu/pams/physics/Physics_Ed/.

Bibliography

[ABPTutorials] *Activity-Based Physics Tutorials*, E. F. Redish, R. N. Steinberg, M. Wittmann, et al. (John Wiley & Sons, to be published).

[Allie 1998] S. Allie, A. Buffler, L. Kaunda, B. Campbell, and F. Lubben, "First-year physics students' perceptions of the quality of experimental measurements," *Int. J. Sci. Educ.* **20,** 447–459 (1998).

[Ambrose 1998] B. S. Ambrose, P. S. Shaffer, R. N. Steinberg, and L. C. McDermott, "An investigation of student understanding of single-slit diffraction and double-slit interference," *Am. J. Phys.* **67,** 146–155 (1998).

[Anderson 1994] R. D. Anderson and C. P. Mitchener, "Research on science teacher education," in *Handbook of Research on Science Teaching and Learning,* D. L. Gabel, ed., (Simon & Schuster Macmillan, New York, 1994), 3–44.

[Anderson 1999] John R. Anderson, *Cognitive Psychology and Its Implications,* 5th ed. (Worth Publishing, 1999).

[Arons 1983] A. Arons, "Student patterns of thinking and reasoning: part I," *Phys. Teach.* (December 1983) 576–581; "Student patterns of thinking and reasoning: part II," (January 1984) 21–26; "Student patterns of thinking and reasoning: part III," 88–93 (February 1984).

[Arons 1990] A. Arons, *A Guide to Introductory Physics Teaching* (John Wiley & Sons, 1990).

[Arons 1994] A. B. Arons, *Homework and Test Questions for Introductory Physics Teaching* (John Wiley & Sons, 1994).

[Ausubel 1978] D. P. Ausubel, *Educational Psychology: A Cognitive View* (International Thomson Publishing, 1978).

[Baddeley 1998] A. Baddeley, *Human Memory: Theory and Practice,* Revised Edition (Allyn & Bacon, 1998).

[Bao 1999] L. Bao, *Dynamics of student modeling: A theory, algorithms, and application to quantum mechanics,* Ph.D. dissertation, University of Maryland, December 1999.

[Bartlett 1932] F. C. Bartlett, *Remembering* (Cambridge University Press, 1932).

[Beichner 1994] R. J. Beichner, "Testing student interpretation of kinematics graphs," *Am. J. Phys.* **62,** 750–762 (1994).

[Beichner 1999] R. Beichner, L. Bernold, E. Burniston, P. Dail, R. Felder, J. Gastineau, M. Gjertsen, and J. Risley, "Case study of the physics component of an integrated curriculum," *Am. J. Phys., PER Suppl.,* **67**:7, S16–S24 (1999).

[Belenky 1986] M. F. Belenky, B. M. Clinchy, N. R. Goldberger, and J. M. Tarule, *Women's Ways of Knowing* (Basic Books, 1986).

[Bligh 1998] D. Bligh, *What's the Use of Lectures* (Intellect, 1998).

[Bloomfield 2001] L. Bloomfield, *How Things Work: The Physics of Everyday Life,* 2nd ed. (John Wiley & Sons, 2001).

[Bowden 1992] J. Bowden, G. Dall'Alba, E. Martin, D. Laurillard, F. Marton, G. Masters, P. Ramsden, A. Stephanou and E. Walsh, "Displacement, velocity and frames of reference: Phenomenographic studies of students' understanding and some implications for teaching and assessment," *Am. J. Phys.* **60,** 262–269 (1992).

[Bransford 1973] J. D. Bransford and M. K. Johnson, "Contextual prerequisites for understanding: Some investigations of comprehension and recall," *J. of Verbal Learning and Verbal Behavior* **11,** 717–726 (1972); "Considerations of some problems of comprehension," in W. Chase, ed., *Visual Information Processing* (Academic Press, 1973).

[Brown 1989] J. S. Brown, A. Collins, and P. Duguid, "Situated cognition and the culture of learning," *Educational Researcher* **18**(1), 32–42 (January–February 1989).

[Campbell 1997] D. K. Campbell, C. M. Elliot, and G. E. Gladding, "Parallel-parking an aircraft carrier: Revising the calculus-based introductory physics sequence at Illinois," *Newsletter of the Forum on Education of the American Physical Society,* Fall 1997.

[Carey 1989] S. Carey, R. Evans, M. Honda, E. Jay, and C. Unger, " 'An experiment is when you try it and see if it works': a study of grade 7 students' understanding of the construction of scientific knowledge," *Int. J. Sci. Ed.* **11,** 514–529 (1989).

[Carpenter 1983] T. P. Carpenter, M. M. Lindquist, W. Mathews, and E. A. Silver, "Results of the third NAEP mathematics assessment: secondary school," *Mathematics Teacher* **76**(9), 652–659 (1983).

[Carroll 1976] L. Carroll, *Sylvie and Bruno* (Garland Publishing, 1976).

[Chi 1981] M. T. H. Chi, P. J. Feltovich, and R. Glaser, "Categorization and representation of physics problems by experts and novices," *Cognitive Science* **5,** 121–152 (1981).

[Christian 2001] W. Christian, *Physlets* (Prentice Hall, 2001).

[Clark 1975] H. Clark and S. Haviland, "Comprehension and the given-new contract," in *Discourse Production and Comprehension,* R. Freedle, ed. (Lawrence Erlbaum, 1975).

[Clement 1989] John J. Clement, "Overcoming students' misconceptions in physics: The role of anchoring intuitions and analogical validity," *Proc. of Second Int. Seminar: Misconceptions and Educ. Strategies in Sci. and Math. III,* J. Novak, ed. (Cornell University, 1987); "Not all preconceptions are misconceptions: Finding 'anchoring conceptions for grounding instruction on students' intuitions," *Int. J. Sci. Ed.* **11** (special issue), 554–565 (1989).

[Clement 1998] John J. Clement, "Expert novice similarities and instruction using analogies," *Int. J. Sci. Ed.* **20,** 1271–1286 (1998).

[Cohen 1983] R. Cohen, B. Eylon, and U. Ganiel, "Potential difference and current in simple electric circuits: A study of students' concepts," *Am. J. Phys.* **51,** 407–412(1983).

[Cummings 1999] K. Cummings, J. Marx, R. Thornton, and D. Kuhl, "Evaluating innovation in studio physics," Phys. Ed. Res. Supplement to *Am. J. Phys.* **67**(7), S38–S45 (1999).

[Cummings 2003] K. Cummings, P. Laws, E. Redish, and P. Cooney, *Understanding Physics,* (John Wiley & Sons, 2004).

[Dennett 1995] Daniel C. Dennett, *Darwin's Dangerous Idea* (Simon & Schuster, 1995).

[Dewey 1929] John Dewey, *The Sources of a Science of Education* (Norton, 1929).

[diSessa 1988] A. A. diSessa, "Knowledge in pieces," in *Constructivism in the Computer Age,* G. Foreman and P. B. Putall, eds. (Lawrence Erlbaum, 1988), 49–70.

[diSessa 1993] A. A. diSessa, "Toward an epistemology of physics," *Cognition and Instruction,* **10,** 105–225 (1993).

[Elby 1999] Andrew Elby, "Helping Physics Students Learn How to Learn," *PER Suppl. Am. J. Phys.* **69(7),** S54–S64 (2001).

[Elby 2001] A. Elby, "Helping physics students learn how to learn," *Am. J. Phys. PER Suppl.* **69**:S1, S54–S64 (2001).

[Ellis 1993] Henry C. Ellis and R. Reed Hunt, *Fundamentals of Cognitive Psychology,* 5th ed. (WCB Brown & Benchmark, 1993).

[Entwistle 1981] Noel Entwistle, *Styles of Integrated Learning and Teaching: An Integrated Outline of Educational Psychology for Students, Teachers, and Lecturers* (John Wiley & Sons, 1981).

[Ericson 1993] K. Ericsson and H. Simon, *Protocol Analysis: Verbal Reports as Data (Revised Edition)* (MIT Press, 1993).

[Flanders 1963] H. Flanders, *Differential Forms, with Applications to the Physical Sciences* (Academic Press, 1963).

[Fripp 2000] J. Fripp, M. Fripp, and D. Fripp, *Speaking of Science: Notable Quotes on Science, Engineering, and the Environment* (LLH Technology Publishing, 2000).

[Frisby 1980] John P. Frisby, *Seeing: Illusion, Brain and Mind* (Oxford University Press, 1980).

[Galileo 1967] Galileo Galilei, *Dialogue Concerning the Two Chief World Systems* (University of California Press, 1967).

[Gardner 1999] Howard Gardner, *Intelligence Reframed: Multiple Intelligences for the 21st Century* (Basic Books, 1999); *Frames of Mind* (Basic Books, 1983).

[Gladding 2001] G. E. Gladding, "Educating in bulk: The introductory physics course revisions at Illinois," talk presented at the Academic Industrial Workshop, Rochester, NY, October 21, 2001.

[Goldberg 1986] F. C. Goldberg and L. C. McDermott,"An investigation of student understanding of the images formed by plane mirrors," *The Physics Teacher* **34,** 472–480 (1986).

[Graham 1996] S. Graham and B. Weiner, "Theories and principles of motivation," in *Handbook of Educational Psychology,* D. C. Berliner and R. C Calfee, eds. (Macmillan, 1996), 63–84.

[HRW6 2001] D. Halliday, R. Resnick, and J. Walker, *Fundamentals of Physics,* 6th ed. (John Wiley & Sons, 2001).

[Hake 1992] Richard Hake, "Socratic pedagogy in the introductory physics laboratory," *The Physics Teacher,* **33** (1992).

[Halloun 1985a] I. A. Halloun and D. Hestenes, "The initial knowledge state of college physics students," *Am. J. Phys.* **53,** 1043–1056 (1985).

[Halloun 1985b] I. Halloun and D. Hestenes, "Common sense concepts about motion," *Am. J. Phys.* **53,** 1056–1065 (1985).

[Halloun 1996] I. Halloun and D. Hestenes, "Interpreting VASS dimensions and profiles," *Science and Education* **7**(6), 553–577 (1998).

[Hammer 1989] D. Hammer, "Two approaches to learning physics," *The Physics Teacher,* **27,** 664–671 (1989).

[Hammer 1996a] David Hammer, "More than misconceptions: Multiple perspectives on student knowledge and reasoning, and an appropriate role for education research," *Am. J. Phys.,* **64,** 1316–1325 (1996).

[Hammer 1996b] David Hammer, "Misconceptions or p-prims: How may alternative perspectives of cognitive structure influence instructional perceptions and intentions?" *Journal of the Learning Sciences,* **5,** 97–127 (1996).

[Hammer 1997] David Hammer, "Discovery learning and discovery teaching," *Cognition and Instruction,* **15,** 485–529 (1997).

[Hammer 2000] David Hammer, "Student resources for learning introductory physics," *Am. J. Phys., PER Suppl.,* **68**(7), S52–S59 (2000).

[Heller 1992] Patricia Heller, Ronald Keith, and Scott Anderson, "Teaching problem solving through cooperative grouping. Part 1: Group versus individual problem solving," *Am. J. Phys.* **60**(7), 627–636 (1992); Patricia Heller, and Mark Hollabaugh, "Teaching problem solving through cooperative grouping. Part 2: Designing problems and structuring groups," *Am. J. Phys.* **60**(7), 637–644 (1992).

[Heller 1996] P. Heller, T. Foster, and K. Heller, "Cooperative group problem solving laboratories for introductory classes," in *The Changing Role of Physics Departments in Modern Universities: Proceedings of the International Conference on Undergraduate Physics Education (ICUPE),* College Park, MD, July 31–Aug. 3, 1996, edited by E. F. Redish and J. S. Rigden, AIP Conf. Proc. **399** (American Institute of Physics, Woodbury NY, 1997), 913–933.

[Heller 1999] P. Heller and K. Heller, *Cooperative Group Problem Solving in Physics,* University of Minnesota preprint, 1999.

[Hestenes 1992a] D. Hestenes, M. Wells and G. Swackhamer, "Force Concept Inventory," *Phys. Teach.* **30,** 141–158 (1992).

[Hestenes 1992b] David Hestenes, "Modeling games in the Newtonian World," *Am. J. Phys.* **60,** 732–748 (1992).

[Hestenes 1992b] D. Hestenes and M. Wells, "A mechanics baseline test," *Phys. Teach.* **30,** 159–166 (1992).

[Huxley 1869] T. H. Huxley, "Anniversary Address of the President," *Geological Society of London, Quarterly Journal* **25,** xxxix–liii (1869).

[Jackson 1998] J. D. Jackson, *Classical Electrodynamics,* 3rd ed. (John Wiley & Sons, 1998).

[Johnson 1993] David W. Johnson, Roger T. Johnson, and Edythe Johnson Holubec, *Circles of Learning: Cooperation in the Classroom,* 4th ed. (Interaction Book Co., 1993).

[Johnston 2001] I. D. Johnston, C. Stewart, R. K. Thornton, and D. E. Kuhl, "The Elusive Right Way to Teach Physics," *AAPT Announcer* **31**(2) 111 (Summer 2001).

[Kanim 1999] S. Kanim, *An investigation of student difficulties in qualitative and quantitative problem solving: Examples from electric circuits and electrostatics,* Ph.D. thesis, University of Washington, 1999.

[Kim 2002] E. Kim and J. Park, "Students do not overcome conceptual difficulties after solving 1000 traditional problems," *Am. J. Phys.* **70**(7), 759–765 (2002).

[Knight 2002] R. D. Knight, *Five Easy Lessons: Strategies for Successful Physics Teaching* (Addison-Wesley, 2002).

[Kolb 1984] David A. Kolb, *Experiential learning: experience as a source of learning and development* (Prentice Hall, 1984).

[Krause 1995] P. A. Krause, P. S. Shaffer, and L. C. McDermott, "Using research on student understanding to guide curriculum development: An example from electricity and magnetism," *AAPT Announcer* 25, 77 (December, 1995).

[Kuhn 1989] D. Kuhn, "Children and adults as intuitive scientists," *Psych. Rev.* **96**(4), 674–689 (1989).

[KuhnT 1970] T. S. Kuhn, *The Structure of Scientific Revolutions,* 2nd ed. (University of Chicago Press, 1970).

[Lakoff 1980] G. Lakoff and M. Johnson, *Metaphors We Live By* (University of Chicago Press, 1980), 70.

[Lave 1991] J. Lave and E. Wenger, *Situated Learning: Legitimate Peripheral Participation* (Cambridge University Press, 1991).

[Laws 1991] P. Laws, "Calculus-based physics without lectures," *Phys. Today*' **44**(12), 24–31 (December 1991).

[Laws 1999] P. Laws, *Workshop Physics: Activity Guide,* 3 vols. (John Wiley & Sons, 1999).

[Lemke 1990] Jay Lemke, *Talking Science: Language, Learning, and Values* (Ablex, 1990).

[Lemke 2000] Jay Lemke, "Multimedia literacy demands of the scientific curriculum," *Linguistics and Education* **10**:3, 247–271 (2000).

[Lin 1982] H. Lin, "Learning physics vs. passing courses," *Phys. Teach.* **20,** 151–157 (1982).

[Linn 1991] M. C. Linn and N. B. Songer, "Cognitive and conceptual change in adolescence," *Am. J. of Educ.,* 379–417 (August 1991).

[Lippmann 2001] R. F. Lippmann, E. F. Redish, and L. J. Lising, "I'm Fine with Having to Think as Long as I Still Get an 'A'," *AAPT Announcer* **31**(2), 84 (Summer 2001).

[Lippmann 2002] R. F. Lippmann and E. F. Redish, "Analyzing student's use of metacognition during laboratory activities." Paper presented at the annual meeting of the American Educational Research Association, New Orleans, LA (April, 2002).

[Loverude 1999] M. E. Loverude, "Investigation of student understanding of hydrostatics and thermal physics and the underlying concepts from mechanics," Ph.D. Thesis, University of Washington, 1999.

[Luetzelschwab 1997] M. Luetzelschwab and P. Laws, *Videopoint* (Lenox Softworks, 1997).

[Ma 1999] Liping Ma, *Knowing and Teaching Elementary Mathematics* (Lawrence Erlbaum Associates, 1999).

[Maloney 1985] D. P. Maloney, "Charged poles," *Physics Education* **20,** 310–316 (1985).

[Maloney 1994] D. P. Maloney, "Research on problem solving," in *Handbook of Research on Science Teaching and Learning,* D. L. Gabel, ed. (Simon & Schuster MacMillan, 1994), 327–354.

[May 2001] D. B. May, "Students' Epistemological Beliefs in an Inquiry-Based Course," *AAPT Announcer* **30**(2), 122 (Summer 2000).

[Mazur 1992] E. Mazur, "Qualitative vs. quantitative thinking: Are we teaching the right thing?", Optics and Photonics News (February 1992).

[Mazur 1997] E. Mazur, *Peer Instruction: A User's Manual* (Prentice Hall, 1997).

[McCloskey 1983] Michael McCloskey, "Naïve theories of motion," in Dedre Gentner and Albert L. Stevens, eds., *Mental Models* (Lawrence Erlbaum, 1983), 299–324.

[McDermott 1991] L. C. McDermott, "Millikan Lecture 1990: What we teach and what is learned—Closing the gap," *Am. J. Phys.* **59,** 301–315 (1991).

[McDermott 1992] L. C. McDermott and P. S. Shaffer, "Research as a guide for curriculum development: An example from introductory electricity. Part I: Investigation of student understanding," *Am. J. Phys.* **60,** 994–1003 (1992); erratum, *ibid.* **61,** 81 (1993).

[McDermott 1994] Lillian C. McDermott, Peter S. Shaffer, and Mark D. Somers, "Research as a guide for teaching introductory mechanics: An illustration in the context of the Atwoods's machine," *Am. J. Phys.* **62,** 46–55 (1994).

[McDermott 1996] L. C. McDermott, et al., *Physics by Inquiry* 2 vols. (John Wiley & Sons, 1996).

[McDermott 1999] L. C. McDermott and E. F. Redish, "Resource Letter PER-1: Physics Education Research," *Am. J. Phys.* **67,** 755–767 (1999).

[McDermott 2000] L. C. McDermott, P. S. Shaffer, and C. P. Constantinou, "Preparing teachers to teach physics and physical science by inquiry," *Phys. Educ.* **35**(6), 411–416 (2000).

[Meltzer 1996] David E. Meltzer and Kandiah Manivannan, "Promoting interactivity in physics lecture classes," *Phys. Teach.* **34,** 72 (February 1996).

[Mestre 1991] J. Mestre, "Learning and Instruction in Pre-College Physical Science," *Physics Today* **44**(9), 56–62 (1991).

[Miller 1956] George A. Miller, "The magical number seven, plus or minus two: Some limits on our capacity for processing information," *Psychological Review* **63,** 81–97 (1956).

[Millikan 1903] R. A. Millikan, *Mechanics Molecular Physics and Heat* (Ginn and Co., 1903).

[Millikan 1950] R. A. Millikan, *The Autobiography of Robert A. Millikan: Portrait of a Life in American Science* (Prentice Hall, 1950).

[Minstrell 1982] J. Minstrell, "Explaining the 'at rest' condition of an object," *Phys. Teach.* **20,** 10–14 (1982).

[Minstrell 1992] J. Minstrell, "Facets of students' knowledge and relevant instruction," In: *Research in Physics Learning: Theoretical Issues and Empirical Studies, Proceedings of an International Workshop,* Bremen, Germany, March 4–8, 1991, edited by R. Duit, F. Goldberg, and H. Niedderer (IPN, Kiel Germany, 1992), 110–128.

[Mitchell 1988] Stephen Mitchell, *Tao Te Ching* (Harper & Row, 1988).

[Moore 1998] T. A. Moore, *Six Ideas That Shaped Physics* (McGraw-Hill, 1998)

[Morse 1994] Robert A. Morse, "The classic method of Mrs. Socrates," *The Physics Teacher* **32,** 276–277 (May 1994).

[Novak 1999] G. M. Novak, E. T. Patterson, A. D. Gavrin, and W. Christian, *Just-in-Time Teaching: Blending Active Learning with Web Technology.* (Prentice Hall, 1999).

[O'Kuma 1999] T. L. O'Kuma, D. P. Maloney, and C. J. Hieggelke, *Ranking Task Exercises in Physics* (Prentice Hall, 1999).

[Perry 1970] W. F. Perry, *Forms of Intellectual and Ethical Development in the College Years* (Holt, Rinehart, & Wilson, 1970).

[Peterson 1989] R. F. Peterson and D. F. Treagust, "Grade-12 Students' Misconceptions of Covalent Bonding and Structure," *J. Chem. Educ.* **66**(6), 459–460 (1989).

[Pfundt 1994] H. Pfundt and R. Duit, *Bibliography: Students' Alternative Frameworks and Science Education,* 4th ed. (IPN Reports-in-Brief, Kiel, Germany, 1994).

[Polya 1945] G. Polya, *How to Solve It* (Princeton University Press, 1945).

[Purcell 1984] Edward M. Purcell, *Electricity and Magnetism* (McGraw-Hill, 1984).

[RedishJ 1993] Janice C. Redish, "Understanding readers," in *Techniques for Technical Communicators,* C. M. Barnum and S. Carliner, eds. (Macmillan, 1993), 14–41.

[Redish 1993] Edward F. Redish and Jack M. Wilson, "Student programming in the introductory physics course: M.U.P.P.E.T.," *Am. J. Phys.*, **61,** 222–232 (1993).

[Redish 1994] Edward F. Redish, "The implication of cognitive studies for teaching physics," *Am. J. Phys.* **62,** 796–803 (1994).

[Redish 1996] E. F. Redish and J. S. Rigden, eds., *The Changing Role of Physics Departments in Modern Universities,* Proc. of the International Conference on Undergraduate Physics Education, College Park, MD, 1996, *AIP Conf. Prof.* **399,** 1175 pages, 2 vols. (AIP, 1997).

[Redish 1997] E. F. Redish, J. M. Saul, and R. N. Steinberg, "On the effectiveness of active-engagement microcomputer-based laboratories," *Am. J. Phys.*, **65,** 45–54 (1997).

[Redish 1998] E. F. Redish, J. M. Saul, and R. N. Steinberg, "Student expectations in introductory physics," *Am. J. Phys.* **66,** 212–224 (1998).

[Redish 1999] E. F. Redish, "Millikan Lecture 1998: Building a science of teaching physics," *Am. J. Phys* **67,** 562–573 (July 1999).

[Redish 2001] E. F. Redish and R. F. Lippmann, "Improving Student Expectations in a Large Lecture Class," *AAPT Announcer* **31**(2), 84 (Summer 2001).

[Reed 1980] M. Reed and B. Simon, *Methods of Mathematical Physics: Functional Analysis* (Academic Press, 1980).

[Reif 1979] F. Reif and M. St. John, "Teaching physicists' thinking skills in the laboratory," *Am. J. Phys.* **47,** 950–957 (1979).

[Reif 1991] F. Reif and J. H. Larkin, "Cognition in scientific and everyday domains: Comparison and learning implications," *J. Res. Sci. Teaching* **28,** 733–760 (1991).

[Reif 1992] F. Reif and S. Allen, "Cognition for interpreting scientific concepts: A study of acceleration," *Cognition and Instruction* **9**(1), 1–44 (1992).

[Reif 1994] F. Reif, "Millikan Lecture 1994: Understanding and teaching important scientific thought processes," *Am. J. Phys.* **63,** 17–32 (1995).

[Reif 1995] F. Reif, *Instructor's Manual to Accompany Understanding Basic Mechanics* (John Wiley & Sons, 1995).

[Rumelhart 1975] D. E. Rumelhart, "Notes on a schema for stories," in *Representation and Understanding,* D. G. Bobrow and A. M. Collins, eds. (Academic Press, 1975), 211–236.

[Sabella 1999] M. Sabella, "Using the context of physics problem solving to evaluate the coherence of student knowledge," Ph.D. thesis, University of Maryland, 1999.

[Sandin 1985] T. R. Sandin, "On not choosing multiple choice," *Am. J. Phys.* **53,** 299–300 (1985).

[Saul 1996] J. M. Saul, private communication, 1996.

[Saul 1997] J. M. Saul and E. F. Redish, "Final Evaluation Report for FIPSE Grant #P116P50026: Evaluation of the Workshop Physics Dissemination Project," University of Maryland preprint, 1997.

[Saul 1998] J. M. Saul, "Beyond problem solving: Evaluating introductory physics courses through the hidden curriculum," Ph.D. thesis, University of Maryland, 1998.

[Scherr 2003] R. Scherr, "An implementation of Physics by Inquiry in a large-enrollment class," *Phys. Teach.* **42** (February, 2003) to be published.

[Schoenfeld 1985] Alan H. Schoenfeld, *Mathematical Problem Solving* (Academic Press, 1985).

[Sere 1993] M.-G. Séré, R. Journeaux, and C. Larcher, "Learning statistical analysis of measurement errors," *Int. J. Sci. Educ.* **15**:4, 427–438 (1993).

[Shaffer 1992] P. S. Shaffer and L. C. McDermott, "Research as a guide for curriculum development: An example from introductory electricity. Part II: Design of an instructional strategy," *Am. J. Phys.* **60,** 1003–1013 (1992).

[Shallice 1988] T. Shallice, *From Neuropsychology to Mental Structure* (Cambridge University Press, 1988).

[Smith 1999] E. E. Smith, "Working Memory," pp. 888–889 in [Wilson 1999].

[Sokoloff 1993] David R. Sokoloff and Ronald K. Thornton, *Tools for Scientific Thinking* (Vernier Software, 1993).

[Sokoloff 1995] D. R. Sokoloff, R. K. Thornton, and P. Laws, *RealTime Physics* (Vernier Software, 1995).

[Sokoloff 1997] D. R. Sokoloff and R. K. Thornton, "Using interactive lecture demonstrations to create an active learning environment," *Phys. Teach.* **35,** 340–347 (1997).

[Sokoloff 2001] D. R. Sokoloff and R. K. Thornton, *Interactive Lecture Demonstrations* (John Wiley & Sons, 2001).

[Songer 1991] N. B. Songer, and M. C. Linn, "How do students' views of science influence knowledge integration?", *Jour. Res. Sci. Teaching* **28**:9, 761–784 (1991).

[Squire 1999] Larry R. Squire and Eric R. Kandel, *Memory: From Mind to Molecules* (Scientific American Library, 1999).

[Steele 1997] C. Steele, "A threat in the air: How stereotypes shape the intellectual identities of women and African Americans," *Am. Psych.* **52,** 613–629 (1997).

[Steinberg 1996] R. N. Steinberg, M. C. Wittmann, and E. F. Redish, "Mathematical tutorials in introductory physics," in [Redish 1996], 1075–1092.

[Steinberg 1997] R. N. Steinberg and M. S. Sabella, "Performance on multiple-choice diagnostics and complementary exam problems," *Phys. Teach.* **35,** 150–155 (1997).

[Steinberg 2001] R. N. Steinberg and K. Donnelly, "PER-based reform at a multicultural institution," *The Physics Teacher* **40,** 108–114 (February 2002).

[Steltzer 2001] T. Stelzer and G. E. Gladding, "The evolution of web-based activities in physics at Illinois," *Newsletter of the Forum on Education of the American Physics Society,* Fall 2001.

[Stipek 1996] D. J. Stipek, "Motivation and instruction," in *Handbook of Educational Psychology,* D. C. Berliner and R. C. Calfee, eds. (Macmillan, 1996), 85–113.

[Suchman 1987] Lucy Suchman, *Plans and Situated Actions: The Problem of Human-Machine Communication* (Cambridge University Press, 1987).

[Thacker 1994] B. Thacker, E. Kim, K. Trefz, and S. M. Lea, "Comparing problem solving performance of physics students in inquiry-based and traditional introductory physics courses," *Am. J. Phys.* **62,** 627–633 (1994).

[Thornton 1990] R. K. Thornton and D. R. Sokoloff, "Learning motion concepts using real-time microcomputer-based laboratory tools," *Am. J. Phys.* **58,** 858–867 (1990).

[Thornton 1996] R. K. Thornton and D. R. Sokoloff, "*RealTime Physics*: Active learning laboratory," in *The Changing Role of Physics Departments in Modern Universities,* E. F. Redish and J. S. Rigden, eds., *AIP Conf. Prof.* **399,** 1101–1118 (1997).

[Thornton 1998] R. K. Thornton and D. R. Sokoloff, "Assessing student learning of Newton's laws: The force and motion conceptual evaluation," *Am. J. Phys.* **66**(4), 228–351 (1998).

[Thornton 2003] R. K. Thornton, D. Kuhl, K. Cummings, and J. Marx, "Comparing the force and motion conceptual evaluation and the force concept inventory," to be published.

[Tobias 1995] S. Tobias, *Overcoming Math Anxiety* (W. W. Norton, 1995).

[Tobias 1997] S. Tobias and J. Raphael, *The Hidden Curriculum: Faculty-Made Tests in Science,* 2 vols. (Plenum, 1997).

[Trowbridge 1995] D. E. Trowbridge and B. Sherwood, *EM Field* (Physics Academic Software, 1995).

[Tutorials 1998] *Tutorials in Introductory Physics,* L. C. McDermott, P. S. Shaffer, and the Physics Education Group at the University of Washington (Prentice Hall, 1998).

[Viennot 2001] L. Viennot, *Reasoning in Physics: The Part of Common Sense* (Kluwer, 2001).

[Vygotsky 1978] L. S. Vygotsky, *Mind in Society: The Development of Higher Psychological Process*, M. Cole, V. John-Steiner, S. Scribner, and E. Souberman, eds. (Harvard University Press, 1978).

[Wason 1966] P. C. Wason, "Reasoning," in *New Horizons in Psychology*, B. M. Foss, ed. (Penguin, Harmondsworth, 1966).

[Wilson 1992] J. M. Wilson and E. F. Redish, "The comprehensive unified physics learning environment: Part I. Background and system operation," *Computers in Physics* **6,** 202–209 (March/April 1992); "The comprehensive unified physics learning environment: Part II. The basis for integrated studies," *ibid.,* 282–286 (May/June, 1992).

[Wilson 1994] J. M. Wilson and E. F. Redish, *The Comprehensive Unified Physics Learning Environment* (Physics Academic Software, 1994).

[Wilson 1999] R. A. Wilson and F. C. Keil, *The MIT Encyclopedia of the Cognitive Sciences* (MIT Press, 1999).

[Wittmann 1998] M. Wittmann, "Making sense of how students come to an understanding of physics: An example from mechanical waves," Ph.D. thesis, University of Maryland, 1998.

[Wittmann 2000] M. Wittmann, "The object coordination class applied to wavepulses: Analyzing student reasoning in wave physics", *Int. J. Sci. Educ.*, to be published (2002).

[Wittmann 2001] M. Wittmann, "RealTime Physics dissemination project: Evaluation at test sites," talk presented at University of Oregon, October 23, 1999, available on the web at http://lrpe.umephy.maine.edu/perl/classes/rtpandild.htm.

Appendix (on Resource CD)

Sample Problems for Homework and Exams
- Estimation Problems
- Multiple-Choice and Short Answer Problems
- Representation Translation Problems
- Ranking Tasks
- Open-Ended Reasoning Problems
- Context-rich Reasoning Problems
- Essay Questions

Action Research Kit
- The Mathematical Modeling Conceptual Evaluation (MMCE)
- Test of Understanding Graphics (TUG-K)
- The Vector Evaluation Test (VET)
- Force Concept Inventory (FCI)
- Force-Motion Concept Evaluation (FMCE)
- The Mechanics Baseline Test (MBT)
- Energy Concept Survey (ECS)
- Heat and Temperature Concept Evaluation (HTCE)
- Wave Diagnostic Test (WDT)
- Conceptual Survey of Electricity and Magnetism (CSEM)
- The Electric Circuits Concept Evaluation (ECCE)
- Determining and Interpreting Resistive Electric Circuits Concept Test (DIRECT)
- Physics Measurement Questionnaire (PMQ)
- The Measurement Uncertainty Quiz (MUQ)
- Maryland Physics Expectations Survey (MPEX)
- The Views about Science Survey (VASS)
- Epistemological Beliefs Assessment for Physics Science Survey (EBAPS)

Bibliographic Resources
- L. C. McDermott and E. F. Redish, “Resource Letter: PER-1: Physics Education Research,” Am. J. Phys. 67, 755-767 (1999)
- L. Jossem, “Resource Letter EPGA-1: The education of physics graduate assistants,” Am. J. Phys. 68, 502-512 (2)

- Useful Books: A list of books that contain discussions of student learning, innovative teaching methods, and interesting problems
- Reading List for a Graduate Seminar in Teaching College Physics for Physicists
- Reading List for a Graduate Seminar in Physics Education Research

Other Resources

- Guidelines and Heuristics: Summary of goals, principles, and commandments
- Writing a Scientific Paper

Resources for Computer Assisted Data Acquisition and Analysis

- MBL information from Vernier
- MBL information from Pasco
- Videopoint demonstration
- WP Excel Tools

INDEX

RCD = item contained on Resource CD associated with this volume.

Credits